ALSO BY ALEXIS MADRIGAL

Powering the Dream: The History and Promise of Green Technology

THE
PACIFIC
CIRCUIT

THE
PACIFIC
CIRCUIT

A GLOBALIZED ACCOUNT
OF THE BATTLE
FOR THE SOUL
OF AN
AMERICAN CITY

Alexis Madrigal

MCD 🌀 FARRAR, STRAUS AND GIROUX

New York

MCD
Farrar, Straus and Giroux
120 Broadway, New York 10271

Title-page and part-opener illustration of Oakland map by
Ink Drop / Shutterstock.

Library of Congress Cataloging-in-Publication Data
Names: Madrigal, Alexis, 1982– author.
Title: The Pacific Circuit : A Globalized Account of the Battle for
  the Soul of an American City / Alexis Madrigal.
Description: First edition. | New York : MCD/Farrar, Straus and Giroux,
  2025. | Includes bibliographical references and index.
Identifiers: LCCN 2024035150 | ISBN 9780374159405 (hardcover)
Subjects: LCSH: Oakland (Calif.)—Economic conditions—21st century. |
  Oakland (Calif.)—Social conditions—21st century. | Pacific Area—
  Commerce.
Classification: LCC HC108.O3 M33 2025 | DDC 330.9794/66—dc23/
  eng/20241003
LC record available at https://lccn.loc.gov/2024035150

Our books may be purchased in bulk for promotional, educational, or business
use. Please contact your local bookseller or the Macmillan Corporate and
Premium Sales Department at 1-800-221-7945, extension 5442, or by email at
MacmillanSpecialMarkets@macmillan.com.

www.mcdbooks.com • www.fsgbooks.com
Follow us on social media at @mcdbooks and @fsgbooks

1   3   5   7   9   10   8   6   4   2

To my family

So we're trapped, trapped, double-trapped, triple-trapped.
Anywhere we go, we find that we're trapped. And every kind of
solution that someone comes up with is just another trap.

—MALCOLM X

Contents

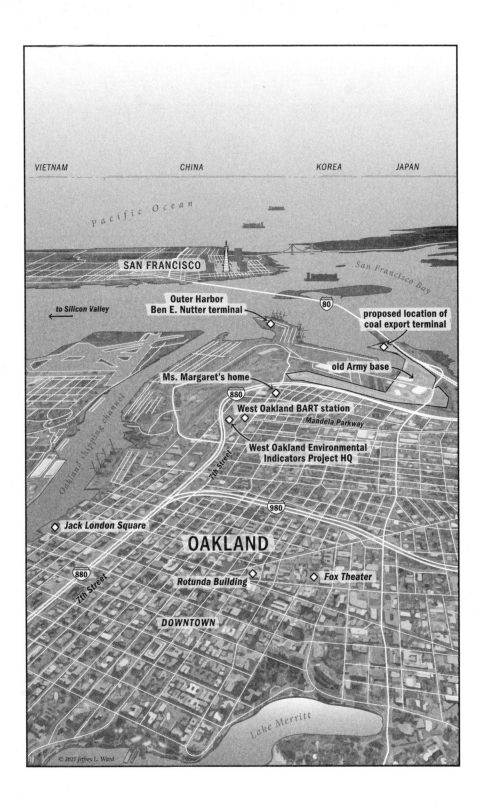

VIETNAM   CHINA   KOREA   JAPAN

Pacific Ocean

SAN FRANCISCO

San Francisco Bay

to Silicon Valley

**Outer Harbor
Ben E. Nutter terminal**

80

**proposed location of
coal export terminal**

old Army base

Ms. Margaret's home

880

West Oakland BART station

Mandela Parkway

7th Street

**West Oakland Environmental
Indicators Project HQ**

Oakland Shipping channel

980

◇ **Jack London Square**

OAKLAND

880

7th Street

Rotunda Building ◇

◇ **Fox Theater**

DOWNTOWN

Lake Merritt

© 2025 Jeffrey L. Ward

THE
PACIFIC
CIRCUIT

Introduction

The whole world at your fingertips,
the ocean at your door.
—BO BURNHAM

What the Mediterranean was for millennia and the Atlantic was for centuries, the Pacific is now. Six of the seven most valuable companies in the world are located along the U.S. Pacific coast, their market values dwarfing the old banking, auto, and oil giants. This is no accident.

The Pacific Circuit is a vast, powerful, opaque cultural structure. It is made up of trade routes and trade deals, human migrations and technological exchanges, coal and oil, cargo ships and corporate relationships. The core network of container ports connects Asian manufacturing and American consumers. These facilities along the water have tendrils that reach far inland to places such as the Inland Port of Dallas, where tens of thousands of logistics workers labor in thousands of acres of warehouses. And it was the Pacific Circuit that opened up small-town America to the flood of Asian-manufactured goods available at the Walmart out by the highway.

The Pacific Circuit has a logic that inheres in the system, pushing for scale and searching for flow. It *does things*. Anytime you open the Amazon app (or the Rakuten app) on your iPhone (or Samsung Galaxy) to order new sponges or a Baby Bjorn, you are tapping into the system, but the same is true when you try out the new pho place or borrow money from a community bank for a house. No city, anywhere in America, has not been transformed by its logistical, technological, and financial mechanisms.

The Pacific Circuit has gotten me, personally, a long way toward understanding what I wanted to know when I started writing this book: What is the relationship between the goods in the containers coming in on the cargo ships and the places where those goods land? Why did the ports develop where and how they did? How different is this new system from previous globalizations, other exchanges between peoples, fallen empires? Why is the rent so damn high?

The United States instigated the creation of the Pacific Circuit. In the wake of World War II, the U.S. government wanted to control the development of Asian countries, keeping them out of the Communist bloc and inside the American sphere of influence. The U.S. propped up right-wing dictators and engineered global trade agreements. We waged two brutal wars, established dozens of military bases around the Pacific, and made countless smaller interventions.

We also made the path to Asian internal development run through American dollars and ports. Their export-oriented economies became tuned to the needs of global markets rather than their own people's conditions. The local operations were often controlled by multinational companies, a shocking number of which were technology companies headquartered in the Bay Area. Americans spread our factories across the Pacific, and put them back together with information. All this to say: we lashed these countries to our own fate, and then thickened those bonds by opening up immigration after nearly a century of making it almost impossible to come.

But that imperial ambition is only the beginning of the arc. Asian countries did not just passively accept American dollars and influence. Japan, Korea, Taiwan, Vietnam, Malaysia, Singapore: these countries developed capacities that no longer even exist in the United States. And then there is China, which emerged in the twenty-first century as a superpower, drawing on and amplifying the Pacific Circuit.

From 1948 to 1990, U.S. imports from Pacific Rim countries increased from $1.5 billion to *$200 billion*. And that was *before* Chinese manufacturing had really spun up. By 2005, imports had topped $500 billion. Finally, in 2022, imports topped $1 trillion to go along with $500 billion of exports.

This complex productive system has an origin: Oakland, California.

So often, the story of a place such as Oakland follows one storyline: Great Migration > urban renewal ("negro removal") > crime and crack > gentrification and displacement. It's told as a story about Black people and white people, largely focused on race relations and government disinvestment. For good reason.

But alongside the more well-known stories about the Black Panthers or Tupac, Oakland became the physical starting point for the Pacific Circuit. Over the years, spreading out from the Bay Area, the whole West Coast has been transformed by the marriage of American capital and corporate know-how with Asian labor and technical capacity.

What makes the Pacific Circuit so fascinating and vexing is that the components are connected in shockingly complex ways, which generate surprising economic and cultural effects. As we'll see, the miniaturization of the transistor, the maximization of the cargo ship, and the securitization of the mortgage are more related than they appear.

There have been consequences to the construction and operation of this economic system. It runs on coal and bunker fuel and diesel, which have created a constellation of local environmental problems spanning many countries. The air pollution is baked in. And, of course, that's the miniature version of the mega problem of this century: global warming caused by the greenhouse gases emitted by modern societies. Perhaps you will not be surprised that the very places that have endured the bad air generated by the operation of ports are *also* those that will be on the front lines of sea level rise. These are the zones that people who ran our cities, gazing upon the potential of Asia, chose to sacrifice.

The Pacific Circuit does not just generate pollution. It also throws off tremendous amounts of cash. Marrying cheaper Asian labor with American product expertise is how our era's great fortunes have been built. But all those dollars have to stop somewhere in their endless circulation, and many of them have stuck in the real estate markets of the United States, most especially in the Bay Area. The median single-family home in Oakland sold for $290,000 in 2012. Ten years later, the median house went for more than $1 million. Every major metropolitan area on the West Coast has a similar set of problems now: housing is too expensive to offset moderately rising incomes for working-class people, so residents are being displaced farther and farther from central cities, often jumping over ritzy suburbs to industrial exurbs.

Another consequence of spiraling housing prices is the unending flow of people falling into homelessness. Whereas contemporary American cities had few informal settlements before 2000, now semipermanent encampments are stitched under highways and along creeks. While the details are complicated in every city and state, the basics are simple: housing has gotten too expensive, incomes have stagnated, and inequality is too great. In 1970, a home in California cost about three times a young couple's annual income. Now it's more than seven times. There are many other statistics that make the point in slightly different ways.

The transformations wrought by the Pacific Circuit extend far beyond economics. Since 1965, the Bay Area has seen an enormous influx of Asian immigrants and their descendants. Asian people are now the single largest racial group in the Bay Area, fully a third of the people. In 1970, roughly 200,000 local residents identified as coming from an Asian country. That number rose more than tenfold, to more than 2 million, by 2020. While other groups have declined steadily (white, Black), or grown erratically (Latinos), the Asian population has simply been a line up and to the right.

The mix of people in the Bay Area, and to a lesser extent in every major West Coast metro, is unprecedented. Of course this is the place that would birth Kamala Harris. Never has there been such an array of people from different countries, united largely by their ancestors' relationship to the U.S. imperial umbrella. Filipinos and Mexicans, Vietnamese and Koreans, Guatemalans and Samoans, Indians and Taiwanese. Our overlapping diasporas—I am part of these movements, too—have made something new.

But it is not only *immigrants* who share this background of migration. Black people, too, were brought to the bay by the Pacific. In a distinct western component of the Great Migration, they came to work in the World War II shipyards and along the postwar industrial waterfront. They settled, and were forced to stay, in precisely the places where the Pacific Circuit touches land: West Oakland, Richmond, Hunters Point, Marin City, East Palo Alto.

←····   ····→

Ms. Margaret is the kind of character you hear tell about, so her reputation should precede her here, too. We will meet her soon enough, but it's worth knowing ahead of time that this book came very close to getting the name *Everybody Knows Ms. Margaret*. Some of it is her style. She likes to *dress*—matching hats and shirts and pants and shoes. Some of it is her other style, which is to say, she is a straight-talking, shit-talking Black woman. She has been known to make people cry, and not only enemies.

If I was going to *really* understand how Oakland bore the costs of the Pacific Circuit, she was the person who could be my guide. Before I really got to know her, I had seen Margaret in action once before. An ally of hers was presenting to a community meeting, and being an eccentric environmentalist white guy, he was . . . floundering. Ms. Margaret popped up from her chair in the front row, grabbed the microphone, and in her halting but powerful way said, "OK, let *me* give you the 4-1-1."

When I was still in the early hazy portion of book research, Margaret would buttonhole people, point to me, and tell them, "He's writing a book about me." Soon enough, it was true. Her life would form the spine of the book, and her mind would inform the way I told the story of her times and her place: the Pacific Circuit, California, the Bay Area, Oakland, West Oakland, Prescott.

In most city politics, there is the official story, the legislative session, the council meeting, or whatever it is. And those things are important. But most everyone knows that's just the thin skin on a much juicier fruit. Ms. Margaret's specialty is understanding those other things, up and down the scale. The report that got written just so someone could point to it at a particular moment in a meeting. The dubious relationship between some business interest or union and some mayor or councilwoman.

I have tried to extend that kind of analysis to the global level, and to incorporate my years as a technology journalist into the story. By working at both hyperlocal and oceanic scales, I hoped to show how Oakland, the town, was being ripped apart by forces far beyond its people's control: technology "startups" from the other side of the bay hollowing out the local economy, financial structures burning through the housing market, ever more trucks rolling through the streets to carry the goods somewhere else.

This place, even though it played a special role in the formation of these systems, could stand in for many other places like it. And as the years went by, I saw what once seemed like uniquely Bay Area problems of inequality, runaway housing prices, homelessness, and precarious gig employment become a standard part of urban life across the whole West Coast and inland cities such as Austin, Denver, and Salt Lake City that have been sucked into our economic vortex.

When the pandemic hit, the Pacific Circuit wobbled, but the cities actually broke. At a moment when they *should* have been preparing for a century of climate disruption and infrastructural rebuilding, Bay Area cities could barely sustain themselves. Violence and drugs spiked in the *very same streets* that crisscross the richest region in the entire world. Despite statistical indicators pointing to a booming economy, few people think the real economy is actually good. Something has gone horribly wrong in the basics of how things work. This book, I hope, helps to explain what happened as the Pacific Circuit wove itself into our aging industrial cities.

The story of the Pacific Circuit, as it emerged from this place, is about the interrelationship of race, technology, American imperialism, climate change, and financial capitalism. This book is an intricate local look at one section along Seventh Street, west of Mandela Parkway, abutting the port, and an intensely global story about how the whole damn system works.

PART I

The State

When drinking water, remember its source.
—VIETNAMESE FOLK SAYING

Who told you the negro deserved a place of refuge? Who told you
that you had that right? Every minute of your life's suffering has
argued otherwise. By every fact of history, it can't exist. This place
must be a delusion, too. Yet here we are. And America, too, is a
delusion, the grandest one of all.
—COLSON WHITEHEAD, *THE UNDERGROUND RAILROAD*

Margaret Gordon

1

Meet Ms. Margaret

Ms. Margaret's office was the former site of an Environmental Protection Agency lead remediation project, and it looked like it: a trailer tucked behind a job-training facility practically in the shadow of the 880 freeway, half a block from Bay Area Rapid Transit (BART).

Inside the West Oakland Environmental Indicators HQ, yellow streamers and shiny ribbons hung from the ceiling. It was Ms. Margaret's birthday. A plastic dining table at the back would soon be filled with Ms. Margaret's friends and family. Past it, there were two small rooms. One was filled with air quality monitoring equipment that they'd trained community members to use. Foam educational posters faded with time and use lined the walls. By the evening, a grandson would end up in this room amid the buzzing equipment, peacefully sleeping in a leather swivel chair, a jacket for a blanket. The devices in that room had been part of Margaret's biggest national recognition, a Champions of Change award from the Obama administration. She went out to the White House and gave a short, nervous speech that I'd watched on YouTube.

Margaret had worn a purple tunic and a big black hat. She'd stepped up to the podium and told her story. "My name is Margaret Gordon and I'm from West Oakland, Oakland, California. I got involved in environmental justice as a second career in dealing with issues from my community caused by the Port of Oakland," she said. "We have trucks, trains, ships, and cargo handling equipment in this neighborhood twenty-four seven. When I moved to the neighborhood twenty-one years ago, I was going to my son's elementary school and I saw a basket of inhalers in the nurse's office. And I wanted to know why there were so many kids with inhalers."

Building on other community efforts, her organization began working with their own air quality sensors and generating their own data in order to get people outside the neighborhood to understand how bad things really were. The data-gathering became a community-building, participatory research project.

That other room at the back of the office was Ms. Margaret's domain alone. Her computer was tucked into the corner. Gumbo bubbled in old slow cookers on a folding table. Ms. Margaret had worked over its components—the meat, the crab, the vegetables, all the seasoning—for ten days, making each bit, then freezing it in preparation for the big day. She had been a cook, even studied culinary arts at Merritt College, so this was a professional at work, but also one who grew up in one of the most southern communities outside the South.

While western and northern, in the racial and California sense, Oakland also has deep links to the western parts of the South. Most Black families in Oakland have links to Louisiana, Arkansas, Oklahoma, and Texas. Margaret's father, Texan by birth, often proudly wore a cowboy hat, though, she noted, sometimes with a dashiki. Many white families have southern roots, too, though they are less likely to celebrate them.

After the official start time, but before the party got going, she tapped a few emails out as she cued up a long YouTube playlist of smooth jazz and R&B. The music crackled out of those old Altec Lansing speakers. She was decked out for her party: a jacket covered with rococo gold swirls over a black Rocawear tunic sweater that was crisscrossed with thick gold chains. Khakis tapered down to her black boots, which were also ornamented with gold.

There was a middle-aged white woman cleaning and bustling around the party. Gloria was her name, I found out. She'd met Margaret in San Francisco, during the years Margaret's family lived in Hunters Point's mixed-race housing projects. They lost touch when Margaret's family moved away. Decades passed. But then they reconnected in Oakland. More decades have passed now. They're now *very* old friends. Gloria took pains to present herself as a simple woman. "I'm not a very verbal person," she told me. "I just cook, clean, do." When Ms. Margaret needed community members to help count the trucks going through their neighborhood, Gloria went to the corner with an engineer's pen and ticked off their trips. When Ms. Margaret needed gumbo reheated and the backroom kept clean, that's what Gloria did. Ms. Margaret hadn't changed a bit over the years, she told me.

"She's a feisty lil woman," she said.

Richard, Gloria's husband, came in. He's an upright kind of guy, tank top undershirt visible under his intensely clean white T-shirt. His body is still strong, but he was stiff at maybe sixty-five. When Margaret needs to get somewhere, Richard often drives her. Young people come and go around her, I would come to learn, but the neighborhood people stayed, if they could.

Over the next several hours, people filtered in. Claudette, her sister, arrived in a layered gray getup, big hoop earrings and rectangular glasses matching her clothes and dyed hair. Like Margaret, she's missing a couple teeth. There was Richard Grow, an energetic, mustached old hand at the EPA. Various elderly white women came through with smallish checks for the organization; 2016 was a good year for that era. They managed to pay Ms. Margaret $17,802. Her co-director, Brian Beveridge, a longtime white resident with more radical politics than you'd expect, got $19,269.

As night fell, the party boomed. Cheap champagne flowed. Boxes of wine, beers. People popped outside to smoke and came in floating. It was New Year's Eve! Oakland was in shock that Donald Trump had just been elected, but people were fired up, too.

The gumbo pots got drained and refilled, drained and refilled. All kinds of people were there: neighbors, friends of Margaret's kids, a young activist named Rain, a Cal professor, and, most prominently, a cadre of members from No Coal in Oakland, an umbrella group that had been fighting the installation of a coal export terminal on what used to be the Oakland Army Base.

The buzz was about a lawsuit that the leaseholder on the land for the terminal, Phil Tagami, had just filed. The city had banned the handling of coal when it got wind of the terminal plan because, of course, a city that voted 96 percent against Donald Trump was not a coal town. These activists, alongside the Sierra Club, which had moved its national headquarters to the city in May 2016, made sure the city council knew this.

The nominal reason for the ban was that the coal trains might spread dust over West Oakland, adding to its already intense environmental burden. But everyone knew climate change was right on the periphery of the argument.

Exporting coal was about the least Oakland thing that anyone could imagine.

And yet Tagami is a native son of the city. He's powerful and rich in a small-city way, and he's got a common touch, despite his three-piece suits and cigars. He'd been behind the showpiece of the New Oakland, the restored Fox Theater. He also made local headlines when, during the Occupy Oakland protests that consumed the city in 2011, he warned off some protesters trying to break into one of his buildings by racking a shotgun. He got a nickname: Shotgun Phil.

He had other names, though: Uncle Phil, for one. His friend Jerry Brown, the governor, declared May 25, 2014, Phil Tagami Day in the state of California. A local vlogger commemorated the occasion with an awkward paean. "The whole point is that you have done a lot for a lot of people," he said. "When you walk into Cafe Van Kleef on a Tuesday night, they call you Uncle Phil because they know you're buying cocktails for the whole doggone bar."

A local muckraker, Gene Hazzard, who speaks at nearly every city council meeting, detained me with long soliloquies about Tagami. Hazzard has made a hobby out of opposing the developer, even selling T-shirts that say "END CITY OF OAKLAND CORRUPTION!!! STOP TAGAMI NOW!!!" He wore the outfit of the investigator, too: a fedora and a scarf, which somehow added to his slight resemblance to Forest Whitaker. He often stopped portentously to direct me to his website for more details. When I checked it later that night, the website felt like a hoarder's digital home. There are links to different news articles and public records scattered haphazardly across the site in a variety of fonts and colors. I'd find out later, too, that Phil Tagami had once backed Hazzard's long-distant (and unsuccessful) runs for city council.

Tagami had been tapped into the power structures of Oakland for decades. Elihu Harris, who governed the city through the '90s, declared a *citywide* Phil Tagami Day in 1993. Later, Tagami was appointed a port commissioner by then-Mayor Jerry Brown. During those days, Tagami liked to hold court at the Fat Lady, a restaurant in Jack London Square that's big with maritime types.

The truth was: everybody loved talking about Phil. He made such a good villain.

Taking some time to show me around, Ms. Margaret walked over to a bookcase by the gumbo room and lugged out a massive three-ring binder. It hit the table with a thud. She pointed to a map of the surrounding neighborhood. Right there, behind her office, was a Superfund site created by the AMCO Chemicals company, which dumped TCE and vinyl chloride into the ground for thirty years. It sat there for twenty-three more before federal funding finally became available to clean it up. As 2016 drew to a close, the EPA was in the midst of installing sixty-nine electrodes in the ground to help extract the toxic chemicals permeating the soil.

The site is not unique. "Seventy-five percent of the properties in West Oakland that are vacant or abandoned have something on them," she said. She opened the binder and pointed to a map. In West Oakland alone, there were 217 toxic sites. I looked at it, horrified. That was one explanation for the emptiness and sadness of the neighborhood outside.

Ms. Margaret, though, exploded with laughter, leaning on her elbows on the table. Pointing to the camera I had slung on my back, she said, "You should have took a picture of your face."

Since I'd moved to Oakland in 2011, I'd often passed through West Oakland, seen empty lots piled with trash, and wondered, what happened here? In less generous moments, I'd even individualize it, and wonder why people were dumping all over the neighborhood. Why weren't other people picking up the trash? But the truth was that the real culprits for all those empty lots were the companies that had set up shop here, shipped the profits to the owners in other neighborhoods, and then skipped town before the environmental bill came due.

The party went on and on. The conversations spilled out of the building. I imagined the scene from above, a drone's-eye view, this trailer next to a Superfund site, surrounded by toxic land, encircled by freeways and train tracks. For all that had been done to this place, for all the global forces hammering at the neighborhood, for all the pressures induced by capital leaving and coming, for all the chemicals that were left here to poison those who stayed, Ms. Margaret and her crew created a glowing center of community and purpose.

Ms. Margaret's territory loops around from her office out to the port and then up Seventh Street past her house and the BART train station. To know Ms. Margaret and to understand the Pacific Circuit, this is the patch of land where you have to start, the dividing line between the city and the port.

I went to take a deep look on a bright winter morning in 2016. The day was crisp. Bright sun, some haze, golden light. California winter. I crossed under the 880 freeway's many ribbons of concrete and found myself in semitruck territory, taking Adeline Street, lofting up over the railroad tracks, past the sign of a long-defunct dockworker bar, the Snug Harbor.

Gaining some height, the industrial landscape came into full view. To the left, Schnitzer Steel extended off toward the shipping channel, which is still generously known as the Oakland Estuary. Great mounds of metal, thirty, forty feet high, hulked in the distance. To the right, train tracks branched and branched, multiplying to twenty across. Squat open hopper cars, tankers, and cabooses sat awaiting locomotion. A bulk world, measured in tons.

The asphalt curved down to Middle Harbor Road, the main drag between the shipping terminals and the tracks. Local kids have been known to make use of the wide, flat asphalt for sideshows, a phenomenon native only to the Bay Area. Cars, burning rubber, drift in circles as the brave and the stupid try to get as close as possible to the hot rods. Everywhere along the road, the rubber remnants of these parties remained. Loop on loop on loop.

The ground beneath my feet was an invention, like so much of the ring of land around the bay from here down to Meta's utopian arc on the muck-pit shore of Menlo Park and up to the old instant shipyards of Richmond. Gold miners fired water cannons at the Sierra foothills, creating torrents of mud and rock and sand. They kept the gold that came tumbling down, and let the rest flow through the massive waterworks of the delta. This is, after all, the largest estuary on the Pacific Coast, a place where the rain trapped by the Sierras and channeled into the San Joaquin and Sacramento rivers goes rushing out to sea.

The mining waste fell where it would somewhere on the floor of the bay. Great dredging machines chomped and slurped up this material, and builders mixed it with whatever else was around, and it became *fill*. Compact it hard enough and it became *land*, new land, histories mixed and buried.

Blindingly white cargo containers rose in stacks opposite the railyard, *Matson* painted on them in blue script. Those steel boxes changed the world, accelerating globalization, but first, they changed Oakland. This place had been an afterthought of an anchorage, a quick pit stop after San Francisco, the great port of the west. But then came Matson, boxing up cargo for Hawaii. And then Sea-Land, which grew into the world's biggest container shipper feeding off the war trade to Vietnam.

Now the whole road is lined with white cranes. They are our town's unofficial logo. In the shipping business, they're called gantry cranes. They straddle the dock on rails placed fifty feet apart, parallel to the water. They can pluck a 50,000-pound container out of a ship like a person swinging up a tote bag.

A big ship might hold 7,500 forty-foot boxes. Full and down, as the expression goes, there might be hundreds of millions of dollars of cargo aboard. Almost everything the world needs to function is right there on those ships. For half a century, the world's companies have created a production system that relies on supply chains. But a supply chain is not a big cable reaching from Asia to California. A supply chain is more like a circulatory system. If corporations are people, the supply chain is how the giants eat. Apple has the most perfect digestive tract in the world. The company has been known to turn over *all its inventory* in less than ten days.

But not much is happening today, it being a holiday. Everything was frozen. No old longshoremen were working the docks, tying up ships. No drayage trucks filled with immigrants eking out a living looping the region's freeways. No massive forklifts flying down the rows of containers, picking just the right box for a trucker to take to the great warehouses of inland California. No man standing in his hot dog shack off the wide shoulder of the road, serving up Gatorade and tamales. No tugboat pushed up on the container ships, captain staring up at the color-block hull rising out of the water like half an arroyo. No dredging barge working its clamshell, chomping away at the bottom of the bay, loading up a scow that will head out somewhere in the ocean and open like a trapdoor, from the center out, sending a grab bag of geological eras plummeting to rest with whale carcasses.

The terminals stretched on for thousands of feet. Two hundred years ago, this all would have been mudflats, a swampy, ecologically important no-man's-land where the streams of the local hills drained to the bay. Now it's all landfill and concrete, except for Middle Harbor Shoreline Park. Environmentalists got this little pocket of land carved in the heart of the port as part of the deal they cut to allow more dredging to accommodate larger ships in the bay. New habitat has been created for birds, who happily scurry forth on tidal flats created from 5.8 million cubic yards of sand and mud that used to sit on the bottom of the now-fifty-feet-deep shipping channel. Climbing up the Chappell Hayes Observation Tower at the park, I could gaze across the bay at San Francisco's gleaming downtown. The still-in-progress Salesforce Tower south of Market newly dominated the skyline. And that building might form the very center of the money fountain that flowed out of the technology industry, coating every neighborhood in the Bay, sending home prices in Oakland soaring.

Next to the park sat the International Maritime Center, where seafarers—the new name for sailors—go for Filipino snacks and religious guidance. The center is out on a piece of land that was reclaimed (what a word) from the bay in the 1960s. It was the last time a big chunk of the tidelands was filled. They used dirt bored out of the hills in the creation of the BART train tunnels. It was there that the first Japanese container ships docked.

The federal government once owned much of the land along the waterfront. These were Navy and Army supply bases, built starting on the day after Pearl Harbor and chugging away through the Pacific wars until they were closed in the 1990s.

In Oakland, the land was turned over to the city and port. The old Army base sat moldering for years, until the new team led by Phil Tagami finally broke ground. In 2016, the future site of Tagami's proposed export terminal remained a forlorn patch of concrete with a million-dollar view of the mothballed original Sea-Land cranes on one side and the new span of the Bay Bridge to San Francisco on the other.

I kept going on Seventh Street and popped back into the city, running right past where BART emerges from its underwater tube. The road dips under the 880 freeway, concrete walls forming a steep canyon. Look at an old map, and you can see that this was the original shoreline, 120 years ago.

I emerged into the west end of West Oakland. The map on my phone called this area the Lower Bottoms. I'd heard some people use the name, seen it scripted on local murals. Lower Bottoms seemed appropriate, but old-timers tended to call it Prescott, after the school that was founded there in 1869.

As I came up to level ground, a white two-story building came into view. ESTHER'S, it read high up in hand-painted red. And down lower, in purple, ORBIT ROOM. JAZZ BLUES MUSIC. People always said this area of town had been a bustling Black commercial district, even before Black migrants arrived during and after World War II. It had really been something, people will tell you, a counterpart to San Francisco's Fillmore to the west or Harlem far to the east.

A city that had been 2 percent Black in 1940 became 47 percent Black by 1980. I knew the general mid-century urban history: Black people were confined to tiny slivers of northern and western cities, usually in the oldest, worst housing, close to the industrial core. As the country subsidized a massive push to suburbanize not just housing but business, Black people were left behind in deteriorating, tax-deprived cities.

The solution that city officials came up with across the country was called *urban renewal*. It was supposed to create new neighborhoods, with better housing and a healthier environment. Looking down Seventh toward downtown, it was clear: nothing had been renewed here. The BART ran screeching overhead. A hulking post office distribution center took up several blocks on the south side of the street. There was nobody around.

Ms. Margaret has lived for decades in a public housing complex called Slim Jenkins Court, right across from the brutalist post office. Right there, in 1967, Huey Newton got into a gunfight with police that left one of the officers dead, Newton wounded, and the Panthers an international cause. In 2016, next door, the Oakland Food Pantry was run out of a couple of trailers that sat on an empty lot filled with planter boxes. This was the site of a potential housing complex for formerly incarcerated people. Its lead developer? The former Black Panther Elaine Brown and her company, Oakland and the World Enterprises. Times change.

A block down, there was the Revolution Cafe. At any given time, it may or may not have been open. The owner, Tony Coleman, ran a nonprofit that helped local kids fix up bikes. They'd taken over the empty lot next door, too, scattering assorted furniture around and filling it with art, most prominently an homage to Oscar Grant, who was killed by police on New Year's Day in 2009.

I kept moving east on Seventh and the homes returned to the south side of the street. The remaining homes are narrow and close together. On the north side of Seventh, extending up through the neighborhood, the old Victorians are bigger, suited for sharing. Tech workers lived in big group houses. It wasn't just Googlers but startup founders, biohackers, cryptocurrency speculators: the weirder, social-justice side of the industry, the people who loved Burning Man and still took drugs for fun.

The last few blocks to the BART stop were mostly deserted—parking lots, shut-in buildings—but there was Upper Kutz, a shop run by a pastor and barber named Truck, and a pupusa joint.

Prescott ends at Mandela Parkway, a wide landscaped boulevard that replaced an oppressive double-decker freeway that had cut the neighborhood off from the rest of Oakland. The local organizer Noni Session, who grew up in its shadow, told me that the day the freeway came down, she walked outside and heard birds singing in the neighborhood for the first time in her life.

On the corner, a housing project rose over a co-op grocery store. Across the street, the West Oakland BART station. Take it one way and you get to downtown Oakland in five minutes. Go the other, and you're in downtown San Francisco in ten.

This part of Oakland has such a specific, powerful history: blues and Panthers, urban renewal and the losses of the crack era, the port and the BART. While there are Black neighborhoods all over this country that share a subset of its stories, Oakland has played a special role in the world, and it concentrates the contradictions of the Bay Area, of California, even of America.

2

Liquid Modernity

Imagine you are wandering through a Target. What would happen if you could reverse the flow of logistics, play it backward? All the lamps and humidifiers and three hundred varieties of deodorant would get stuffed back into containers, trucked back to distribution centers out in the exurbs of the Bay Area, then sent back down to the port on a truck, and finally loaded back onto a ship, which would cross the ocean to an Asian port, and trucks and trains there would head back to the factories whence they came. Keep going and you'd see the raw materials flying out of the products back to the forests and the underground and the feedlots and the farms and the industrial chemistry facilities and the oil fields. Literally anything on the shelf is part of this extravagantly complex, energy-intensive system. This incredible network is the main loop of the Pacific Circuit, and it is also Target, the number two importer in America, and the place where everybody goes in the Bay. And it is also Walmart, the largest private employer in the United States. And it is also Amazon, the second-largest private employer. The field of logistics escaped the military and went straight into the heart of American retail.

At each step, there's a beep, just like at the checkout counter. A barcode is being read and that data fed back into the systems that control these assemblages of products. Everything you buy, like every container out at the port, exists both as itself and also as a digital manifestation that lets the logistical systems see it and move it around.

Though the avatar of the Pacific Circuit would have to be the container or the crane, it could not exist without all the digital components.

The jobs at the docks are subject to the same tools and logic of high technology as those in the San Francisco towers south of Market.

In the early 1970s, these emergent systems of production were being called "total logistics" in industry trade journals because each product could be moved according to its "physical characteristics, transportation patterns, and control functions." Optimize the variables, harmonize the rhythms, and massive efficiency would result. Any system could be modeled as a series of outputs and feedback loops, inputs and feed-forward circuits, containers and operations. Men, a 300-ton stamping press, the tides, a mineral in West Africa, the sleep schedules of Taiwanese workers, planes, trains, trucks, ships.

The shipping container made all this vastly simpler. Each container is exactly like millions of other containers in its basic form and capacities, perfectly standardized through agonizing protocol meetings. The internet packet, the unit of the basic protocol for moving data around the world, is an equally powerful abstraction. Together, they form the one-two punch of the modern world.

I have always loved this diagram from an early military document about containerization. It's what just about everyone would recognize as a container, and it is marked "FIGURE 1. CONTAINER." It might seem silly at first, but it is the very emptiness of the drawing that makes the point: a container is important because anything can travel inside one, and you don't need to know anything about it except one thing: that it is exactly like the other containers.

But, *but* . . . unlike internet packets, containers are also made of steel and can weigh two tons. Moving a container requires a truck, a big diesel truck. And a ship, a huge ship. And massive machinery along the dock.

"The overriding feature of these logistical worlds is the contradiction or the tension between the impulse and desire to dematerialize everything," the scholar Jesse LeCavalier told me, "and then the stubborn reality of having to actually move it around."

Shipping companies figured out how to deal with the tension: they scaled. As transistors got tinier and tinier, fueling Silicon Valley and the growth of the technology industry, containerized shipping moved in the opposite direction, growing and growing, fueling global trade. Ships outgrew the older cranes, the older docks. Even the canals humans had dug to connect the world's oceans had to be widened to accommodate the girth of the newest ships. The unit cost of shipping a container fell and fell.

The miniaturization of technology and the gigantism of logistics are of one piece. Herb Mills, perhaps the most incisive of all the longshore philosophers—and a credentialed political scientist—saw the dual transformation firsthand. "Containerization is the technological underpinning of the global economy. That and . . . computers, rapid communication, inventory control, and all that stuff—is why jobs have flowed out of the United States. Because you can bet your sweet ass if all them transmissions was being hand-handled out of the hold of a ship and put on a pallet board and sent ashore, rather than twenty tons of transmissions being in a goddamn container box, transmissions would still be built in Detroit," Mills said. "So the container has been the physical means of exploiting cheap labor throughout the world."

And what was true for transmissions is also true for basically everything at Target. In 2004, a cargo ship got caught in a storm near Monterey, California, and dropped some of its containers into a marine sanctuary. The subsequent investigation revealed a list of the contents of the fallen boxes: 594 kennel kits of cyclone fencing; 2,260 Michelin tires; 132 hospital beds (Dalton Model B-2000); 2,492 mattress pads; 61,200 hair ribbons, 12,000 hair wraps, 25,500 turbans, and 25,920 barrettes; a variety of leather furniture and 479 seat cushions; and 38,000 pounds of waste cardboard.

Each of these is also just a box, a number, a line in a database or in many databases. Container GATU8496935, the hair products box, is now positioned deep in the Pacific Ocean, home to crabs, rusting. But that is the exception.

More than a million boxes will pass through Oakland, imports and exports. Three quarters of these boxes go to or come from Asia. The same is true down at the even bigger ports in LA/Long Beach and up in Puget Sound. The scale of it all is sublime, in that sense of awe that borders on terror.

There's essentially no overstating the importance of this planetary system. And if you need more evidence, you can see how the economy cracked and inflated when it began to malfunction *even just a bit* during the peak of the pandemic and early years of the Biden administration.

Over and over, for these past ten years, I would hear the Polish philosopher Zygmunt Bauman in my head, his theory of "liquid modernity." Summed up in one phrase: we live in a time when *anything can happen, but nothing can be done.* Had the years since 2016, the years I've been working on this book, proved anything more than this? "Living under liquid modern conditions can be compared to walking in a minefield," Bauman wrote. "Everyone knows an explosion might happen at any moment and in any place, but no one knows when the moment will come and where the place will be."

Trumpism, COVID, supply chain snarls. The institutions one might have imagined would be there to do something turned out to be able to do . . . almost nothing. And a big part of the reason is simply that countries, even big, powerful ones such as the United States, do not control the functioning of this system, at least not completely.

There are continuities between our times and those that preceded us, but power has moved. It no longer resides locally. Often, it is not human. Local and national borders don't define it neatly. Power is as liquid as the Pacific.

And yet we do all live in a place. And those places shape how we see the world. Oakland is Boots Riley's surrealist labor territory, literally these blocks. Jenny Odell goes birding out here. So many other Oakland artists and writers and thinkers have opened my eyes to different aspects of things: Liam O'Donoghue's brilliant histories, the public health research of Tony Iton, the photographer Demondre Ward's street portraits, my KQED colleague Pendarvis Harshaw's love letters to the town, the librarians Betty Marvin and Dorothy Lazard, Rafael Casal and Daveed Diggs and Tommy Orange. That's not to mention the anarchists and Burner weirdos, radical biotechologists and SoundCloud rappers, militant botanists and indigenous restaurateurs, dance crews and community archivists. Oakland people are just different.

We don't agree on everything, but I do think living here suggests a certain deep Oaklandish politics. Oaklandish, technically, is a clothing brand here in town. They once made a shirt showing a cargo ship stacked with containers, which throws up a shadow of the city skyline. I'm not talking about the brand, per se, but you see what I am saying about this place? We attend to the back end of industrial society. The first thing most people see of our city is the port and its massive cranes, containers stacked in the yards. Infrastructure is alive here, not just as technological sublime of steel and concrete but cut through with the labor of individuals doing all the things that connect us physically to the production of the global economy.

Oakland is polyglot, multiethnic, multiracial. While for brief periods, one demographic group held sway, Oakland has generally been a mix of people. Oaklandish people are highly conscious of the ways that institutional discrimination has structured every aspect of American society—and also of the tremendous cultural ferment these circumstances have produced. The pain and pleasure of belonging and not belonging, of longing for some other home and making a new one, of forgetting and of remembering.

While San Francisco and Silicon Valley have dedicated themselves to the new, to the future, Oakland is not so sure. Ambivalence is a respectable position to take about new technologies. Oaklandish people want to pick and choose what innovations structure their lives. We want to shape some of them to our ends and leave other ones alone. Our proximity to the fountainhead of technology means that we are deeply familiar with the new shit, but we are not fully inside the logic of its producers. Our futurism is a necessity, not a luxury. Oaklandish futurism has to be intersectional. When, exactly, would the 70 percent of Oakland residents who are Latino, African American, or Asian be better off in these United States of America?

Oaklandish people are looking around and listening to the place itself. Some of this is not by choice. The bay and hills define the city seismically, meteorologically, economically. The landscape is impossible to ignore. It's so beautiful that anytime I'm high up in a building or on a freeway overpass or on a hike in the hills, I can be stunned into silence by a glance at the land and water. They are also our connection to the original stewards of this place, the Ohlone people, and to the internal processes of the Earth, over which all of life is but the thinnest film over deep time.

But in parts of Oakland, the history embedded in the ground in the form of pollution can hurt you. The air gives your kids asthma. The water is not for swimming. But what beauty and possibility in the degraded landscape! Healing these places unlocks civic space, giving life. A perspective level up, the whole planet, climactically and otherwise, is a degraded place. This is where we live. There is no virtue in setting out to find a pristine wilderness or live apart from *this* world.

Oaklandish people, in so many guises, are trying to create collective space and action. Cooperatives, gardens, event series, community archives, dance crews, homeless camps, ad hoc committees. When everything in the economic order is trying to blast your self into monetizable particles staring at a phone, the Oaklandish solution is to ground down into the place, its people, the flows of labor that move everything, the historical guide wires that structure, dampen, and amplify the present.

So let's start with the most important fact about container shipping: bigger is better. The more boxes that can be packed on a ship, the lower the cost for any single box. So the ships have gotten bigger, and that cold fact has driven so much else: the rising height of cranes, the bigger pulses of trucks into ports, the expanding area needed to store containers, the growing consolidation of trade routes. These big ships are ever more expensive, which means that each vessel requires a larger chunk of capital. Few companies can pull together that money. So the top four shipping lines now control more than half the market, and three big "alliances" have a hold on roughly 85 percent of it.

The same has happened with ports. The biggest ports are getting bigger, the rest are struggling. Oakland's is in the middle, almost certainly doomed to decreasing relevance as other, larger ports keep winning.

For Oakland people, that may be surprising. We put cargo ships on T-shirts and see them as part of our landscape, same as night herons or redwoods. And there's no point in denying that these ships are beautiful, at rest, where their gently ravaged hulls look like Rothkos, or slowly vectoring out the Golden Gate, tugboat escorts arrayed around them.

Alongside *that* reality, there is another one: the circuit requires sacrifice zones, places that have to bear the costs and pain of storing and moving goods, harnessing fossil fuel and transportation infrastructure. It has been this way since the dawn of containerization.

Look at the places that became the first major container ports in the United States: Newark and Oakland. Why these places? They had waterfront land, natural logistics links, and . . . the residential areas adjacent to the ports were filled with poor Black people who were marginalized by the political system. Why did those people live in those areas, close to the sewage reek of the shoreline, breathing air fouled with fumes from heavy industry, adjacent to military facilities, surrounded by freeways, bereft of open space?

The land people lived on had been layered over with abstractions—property, race, risk. Racism, in this sense, was not some ambient property of human relations but a set of very specific ideas, laws, bureaucracies, and quantitative evaluations. A new style of segregation swept across the country with the New Deal. It shaped who could live where, what kinds of pollutants would enter your lungs, where your kids would go to school, and—even in our present day—whether you could plant vegetables in the yard actually expecting to eat them.

We have to start with this history, not because the solutions to our problems are hiding back there but because otherwise nothing in West Oakland makes sense. If the ports of the emerging Pacific Circuit required lots of waterfront land adjacent to cheap real estate, why was the real estate worth what it was worth?

3

Under All Is the Land

Believe it or not, an important component of the American real estate system can be traced to a guy named Frederic Clements crisscrossing the American grasslands, surveying plants for a publication, *The Phytogeography of Nebraska*. In the late nineteenth century, the scientific world was awash with exciting theories about how reality worked. Geology, biology, physics. Earth was millions of years old. Sea creatures had once lived on mountaintops. The eons underpinning the development of every point on the planet required new ideas.

The first and most lasting intellectual payoff of this era was Charles Darwin's *On the Origin of Species*. The book's theories exerted a strong influence on Clements. Behind the teeming, screeching, swimming, flying, single-celled, thinking diversity of life, there was a set of rules that could explain how it all came to be. Fred Clements took on the task of figuring out how all the plants in the world came to be where they are. From Nebraska to the globe.

Ecologists such as Clements and the Danish father of the field, Eugenius Warming, sensed deep laws at work. Where others saw individual seeds sprouting or specific trees falling in the woods, they saw holism and order. Clements argued that groups of plant species snapped together into superorganisms he called formations. These developed in predictable, uniform ways. The conditions of a place served as a climactic "genome," as a later scholar would put it, which orchestrated the development of the flora toward an inevitable "climax" formation composed of the plants that *should* rule a region.

A bog plant might create the conditions for a bullrush, a moss on a rock for a shrub. This was the way of plant life on Earth, Clements argued, until climax. Then that formation's natural dominance ended the sequence. Only an outside disturbance—fire, human development, flood—could reset the clock, and send the area's plants working back toward destiny.

The theory of succession, as this came to be called, became part of the rootstock of modern ecology, and though it would eventually fall to mounting evidence, Clementsian ecology slipped out of the discipline and became a fundamental way of seeing the world for an influential group of social scientists.

The University of Chicago had become a key center for urban sociology, and its defining document was a 1,040-page course pack, an anthology of ideas that could be applied to peoples and societies. Adam Smith shares space with Helen Keller, Francis Bacon with Charles Darwin, naturalist with sociologist, philosopher with eugenicist. And hey, there is Fred Clements.

The *idea of succession*, in fact, turned out to be the engine of their theory of the city. "In the expansion of the city a process of distribution takes place which sifts and sorts and relocates individuals and groups by residence and occupation," Chicago's Ernest Burgess wrote. "The resulting differentiation of the cosmopolitan American city into areas is typically all from one pattern." One pattern!

That pattern was concentric rings. In the center rose the business district, the Loop, as it was known in Chicago. A "zone in transition" encircled the core, a mix of business and manufacturing that they sometimes called the "area of deterioration." Workingmen lived beyond that, themselves surrounded by a ring of nicer housing of "restricted" residential districts. And the final ring they called the "commuter zone."

The city had a logic. It grew from the center outward, due "partly to business and industrial pressure and partly to residential pull." Shaping that general movement were "natural economic and cultural groupings." Racial, ethnic, and class groups found their "natural area" within the city as well as occupations that suited their "racial temperament," which explained "Irish policemen, Greek ice-cream parlors, Chinese laundries" as well as "Negro porters, Belgian janitors, etc."

Burgess published a diagram that included the racial and cultural dimensions of urban growth. It showed the (Jewish) Ghetto, Little Sicily, Chinatown, the Black Belt, and Deutschland. "This chart brings out clearly the main fact of expansion, namely, the tendency of each inner zone to extend its area by the invasion of the next outer zone," he wrote. "This aspect of expansion may be called succession, a process which has been studied in detail in plant ecology."

In the human world, succession had three steps, according to Burgess. First, there was an "invasion," as one cultural group moved into another's territory on its inevitable outward trajectory. The existing residents of that zone of the city would engage in *reaction*, "the resistance mild or violent" to the newcomers. Nonetheless, in most cases, a "rapid abandonment of the area by its old-time residents" would result from the influx of newcomers.

The neighborhood would settle into a new, degraded state. It's worth noting that there was no possibility that an "invasion" would end in a stable mixed neighborhood. The process, once begun, evolved in a predictable direction. "Race prejudice alone" was not "in the main" to blame for the extreme segregation found in northern cities, Burgess wrote, but "the result of the interplay of factors in urban growth which determine the location and movement of all groups, institutions and individuals."

Clements's theory about plants in their environments became a guide to understanding and *predicting* human behavior in that most complex configuration, the city. The human ecologists naturalized the racism that Black people experienced, north and south. It just *was*, a force of nature, like electricity. This theory sanctioned the open discrimination of the real estate industry and among city planners as acting in accordance with the latest social science. It guided the decisions of appraisers, and even filtered down to the Irish policeman or Belgian janitor seeing a Black family move in next door. Burgess's argument was not merely an observation that this had happened in the past, under particular racial hierarchies and in response to specific patterns of migration or historical oppression; this was a predictive theory about *how cities worked*, generally. City planners or government officials could not counteract these forces any more than an architect could ignore gravity.

I sometimes admire the ambition of the Chicago School people. What a task to set before yourself: figure out the city! Imagine looking out at Chicago—the archetypal industrial city, built from the bodies of animals, conflagration, coal, power; a glorious, brutal, polluted, fractious place—and seeing *order* and *reason* and *science*. Imagine the power of that kind of deep knowledge. The sociologists could see evidence for the succession theory all over the country, too. Ethnic and racial fights over city territory happened all the time. Wasn't it better to ground them in some universal principle rather than a debased human morality or the failings of one ethnic group or another?

Burgess's key collaborator, Robert Park, was, he would have assured us, a friend to the negro. Park had worked as a reporter, where he'd exposed some of the more hideous atrocities of the Belgian colonial system in the Congo. Later, he'd abandon his family for long stretches to work for Booker T. Washington at the Tuskegee Institute. After his retirement, he taught at the predominantly Black college Fisk University. He coined the phrase "marginal man," which is still in common use in the sense of "marginalized" communities.

As president of the American Sociological Association, he gave a speech at the annual conference that became a landmark in the field. It was where they unveiled their theory of Chicago and every other city. They called their breakthrough approach "human ecology," and it relied on the Chicago hypothesis that "social phenomena are subject to mathematical measurement" because "social relations are frequently correlated with spatial relations." *Where* things were happening told you *what* was happening. The marginal man was marginal *because* he was literally on the margins of things.

It's worth dwelling, for a moment, on how hand-wavey the model was. The book *American Apartheid* demolishes the notion that what Black people were experiencing was analogous to other groups. There might have been neighborhoods called "Little Italy," but the vast majority of Italians did not live in those places. Black people were far more segregated and isolated than the Irish, Italians, Poles, or any other group—and by vast margins. The basics of the theory were floppy, too. Even supporters had to admit that cities did not *actually* grow in rings.

But there was another key problem in this scientific endeavor, beyond the thin real-world evidence that their theories had. Because we're talking about people here, and not lichens, their ideas could actually enact themselves. Their model, to borrow the contemporary sociologist Donald MacKenzie's phrase, was "an engine, not a camera." The more people came to use and believe their model, the more reality would begin to conform to it. But to do so, the theory would have to leave the academy and become part of the laws and processes of the nation.

Well, it did.

Human ecology hybridized with emerging ideas about quantification, mapping, and other ways of managing the modern world to remake American cities over the subsequent forty years, tearing apart the urban sanctuaries that Black people had built while providing intellectual cover for undeniably racist policies.

People trained in the Chicago School methods of evaluating and understanding cities entered real estate appraisal, mortgage financing, and every level and branch of government, from the workaday employees at the Federal Housing Administration to the Supreme Court. These theories, hammered into policies and procedures, reshaped the opportunities available to generations of African Americans, and whoever else lived beside them. It's why, as you read Margaret Gordon's story, it will share so many features with those of hundreds of thousands of other families across the country. A whole system of thought grew up around a partial, not-quite-correct understanding of how plants interacted with one another. That filtered into the planning processes of the federal government, and then was executed by the (white) power structures of American cities as the nation reeled from the Depression and World War II.

The crucial vehicle would be the government's interventions in the financing of residential real estate. Racial group movement became a formalized element of financial risk calculations, based on the Chicago School's conclusions. In building the American mortgage, one of the world's most important financial technologies from then to the present, our government built the American middle class, and also eviscerated the inner city.

←···   ···→

In February 1934, the Oakland Real Estate Board gathered at the Hotel Leamington for a luncheon talk by Chicago's Frederick Babcock, "authority on development trends and real estate appraising." A prominent author of books on the subject, Babcock had become "well known to Oakland realtors," according to the *Oakland Tribune*. Imagine that room. A ballroom, chandeliers. Long banquet tables. White cloths. Rolls, pads of butter. Young upstarts angling for a good spot among the more established real estate elders. All men, all white men. Coat-and-tie, certainly, perhaps double-breasted. A moment for connections to be made and common knowledge to be forged.

In October, Babcock came back as a newly minted Federal Housing Administration official to discuss the "Risk Rating" of loans. The FHA was a linchpin of FDR's New Deal. After the spectacular collapse of real estate markets during the Depression, the FHA stepped in to stabilize the market and shoulder some of the risk for lenders. Banks could make loans knowing that the federal government would backstop them. This mortgage insurance proved crucial to the expansion of American homeownership, especially at the lower end of the middle class. If a loan fit into the rubric dictated by the FHA in its handbooks, it "conformed," and could get FHA insurance, lubricating the loan for the bank. So what the FHA wanted is what lenders ended up doing.

What did the FHA want? Babcock's plan was to make "an incisive analysis of hazard," what we would now call risk. If you were gonna lend money long-term (say, thirty years), then you had to know both the present state of a property and how its value might change. This would require "the collection and tabulation of much statistical data not now generally available, and a request is being made for the resumption of the national real estate inventory," the *Tribune* reported.

Over the next two years, Babcock would develop this system at the

FHA while dozens of Oakland real estate brokers would be trained in Babcockian appraisal methods through months of classes. They commingled with people who'd gone to work for another new division of the federal government, the Home Owners' Loan Corporation, or for Oakland's city planners. Locals would help the federal government see and risk-rate all of Oakland. Its neighborhoods would be statistically graded, so that the government could decide what areas it would help grow and what areas it would allow to complete their succession-driven decline. These maps would become famous and controversial, and these practices would get a name: redlining. But that label doesn't come close to describing the variety and intensity of the ways people at every level of government *and* in the real estate industry worked against marginalized people generally, and Black people specifically.

The FHA created "underwriting grids" with a complicated scoring system that worked like an algorithm. Each borrower, property, and property *location* received a rating that was, itself, composed of a variety of factors. Within the location rubric, there were nine factors, ranging from "Protection from Adverse Influences," the most important, to less important factors like "Sufficiency of Utilities and Conveniences." An adverse influence, of course, could be an "inharmonious" racial group or, say, a copper smelter.

While the scoring system might sound complicated, and presaged the kind of algorithmic decision-making that has come under intense scrutiny in recent years, the unmistakable stank of humanity remained on the card. Government officials didn't have to complete the whole process if they decided that the risk of any one component was too high. They could just write "REJECT" on the total rating and be done with it. That rejection meant no FHA insurance for that particular loan, which would, at the very least, make it more expensive, and often was a deal killer.

As Babcock had indicated, there were some problems with developing their risk-rating system, namely that they didn't know enough about the places they were trying to evaluate. They needed a layer of information, since the federal government was not going to be intimately acquainted with every area in the way a local banker might be. So the

New Dealers incentivized cities to do "Real Property Surveys." They could even get labor help from the Works Progress Administration.

In Oakland, I. S. Shattuck, then the city planning engineer, directed this substantial task, which took most of the fall of 1936. Shattuck held that the importance of the survey "can scarcely be over-emphasized." Housing was at the root of many of the issues of a growing city, and "to remedy existing conditions, definite and accurate information must, first of all, be made available."

Data was amassed on *every single block* of the city. Oakland created two sets of maps on which it placed the most important information. Blocks were shaded yellow, orange, red, green, and blue to indicate (from lowest to highest) the average monthly rental price of housing on that block. Then, in tiny numerals, seven other pieces of data were recorded: a bunch of stuff about the buildings themselves *and* the percentage of families that are non-white ("Negro" or "Oriental"). There were also stacks of maps made from individual data points plotted onto an Oakland base map, precisely as Chicago sociologists had promoted doing.

This was a remarkable tool for doing work. You could get a synoptic view of the city's expensive (good) and cheap (bad) neighborhoods. You could see trends and arterials. But then you could look more closely at a set of blocks and check out the details on any of them. In a time before computing, before Excel spreadsheets and Google Maps, this was state-of-the-art spatial technology.

The report Shattuck wrote from this data fully bought into the Chicago ideas about urban growth and decay. "Oakland has not yet felt extensively the dire results of ordinary city growth carrying in its wake depreciated districts which gradually drop into the 'Slum' classification or the 'Blighted District' classification," he wrote. "There are of course some districts where it will be obvious after further intensive housing study that only severe measures can be successfully applied, such as demolition and rebuilding."

"Some districts" meant West Oakland, particularly the racially mixed western part of it. This area, according to the principles of sound urban planning at the time, had to be segregated. As the western reaches of West Oakland had been zoned for heavy industry, despite (or because of?) their large, multiracial residential populations, Shattuck was sure that West Oakland's population would decrease as industry made the air and land increasingly inhospitable. This "crowding out" would help get rid of some of Oakland's bad housing, but he admitted, "living conditions in the dwellings remaining will certainly fall further and further below a decent standard."

The city fully understood that it was letting heavy industry choke and poison residents. Pollution as policy. The point was to create a neighborhood so degraded that no one would want to live there. Anything remaining: demolish and rebuild.

But that was not Shattuck's only plan. He also suggested that a freeway be cut right through the heart of West Oakland. "The housing advantage of this diagonal highway route is that it would segregate that portion of the district under discussion where non-white families are largely concentrated from the portion where they are relatively few in number," Shattuck argued.

The pollution and segregation of Black neighborhoods was a choice that white planners made.

Somewhere in Oakland sometime in that fall of 1937, two men named Ralph filled out a form for a federal government agency, the Home Owners' Loan Corporation, noting a problem with an area of Oakland that was labeled D-5: "Infiltration of colored residents. There are now about twelve families scattered over the area indicated."

Ralph E. Prentice and Ralph Spencer, the two local real estate men, provided "clarifying remarks" to help others understand why they'd labeled the area as they had. "Unless one knows about the colored families living in the district, there is no means of distinguishing their homes from those of their white neighbors," the Ralphs wrote. "The homes of the Negroes are in many instances better kept than the adjoining homes of white owners. Loans in this area should be governed according to hazard." The neighborhood would have to degrade, and yet there were the "better kept" houses of the infiltrators, sitting there as a rebuke to the theory of decline baked into succession.

Something about this document hit me. The specificity of that *twelve*. This knowable number of Black families had literally changed the map produced by the federal government about this region, channeling financial support away from the area.

These maps became famous thanks to the work of the historian Kenneth Jackson and his book *Crabgrass Frontier*. I came upon the D-5 document while I was working at *The Atlantic* and Ta-Nehisi Coates was preparing to release his landmark essay, "The Case for Reparations." Along with my colleagues, we were preparing to write some stories that would run as a package online around the essay. So I started looking into my own neighborhood in North Oakland, right by the Berkeley border, and a block from D-5. In absolute terms, this isn't far from West Oakland. It's just a few miles, but then and now, these are very different places.

While most of the red areas on the map were contiguous with each other, this particular one was alone, a splotch of red. There were no other "adverse influences." It really was only the twelve Black families that got this area marked as high-risk and probably on the downgrade, despite the efforts of the homeowners.

Who were these people, I wondered? I went looking in the census records, and almost immediately noticed one large family living in a house on Telegraph Avenue. They stuck out because one of the residents, Archie Williams, was an aviator. It turned out that was the least of his accomplishments. Williams won a gold medal at the 1936 Olympics—the Hitler Olympics! When he returned to Oakland, the headline in the *Tribune* blared, "He's Back—Oakland Pays Him Homage Due a Champ." He got the literal keys to the city handed to him by the mayor and rode on a fire engine in a parade from the 16th Street railroad station through West Oakland to downtown. His grandmother greeted him, crying with pride and happiness. The Junior Chamber of Commerce gave him a gold watch. Standing in front of the crowd of eight hundred people at City Hall, who cheered him on as the hero he was, he was moved to local affection, at least as reported. "I have traveled all over the United States and Europe in the last three months," Williams said, "but right now I would trade them all for this little spot in my home town!"

Williams was born in Oakland in 1915, and his father died when he was ten. His mother worked as a domestic servant for a white woman in San Francisco, so he mostly saw her on the weekends. He spent most of his time with his extended family, several of whom lived together in that house on Telegraph. His grandpa was a veteran of the Spanish-American War. His uncle was an embalmer. His grandmother Fannie Wall was an activist, "a race woman," as the Black women in civic clubs were known. Wall helped co-found the only orphanage that accepted Black children in all of California, down on Linden Street and Eighth, just a few blocks from where Margaret Gordon would live two generations later.

They were an NAACP household. The first magazine Archie ever saw was its publication, *The Crisis*. As an old man, he still remembered the photographs of lynchings that ran in its pages. Williams's family was upwardly mobile—folks who bought into the idea embodied in the motto of the National Association of Colored Women: "Lifting as We Climb." And the family had climbed way up out of the port, all the way to the northern pole of Oakland, and halfway up the physical and socioeconomic gradient that ran from the shore to the hills. When Archie walked outside the house, he would see not the railroad tracks or the finger piers but the campanile of the University of California, where he studied engineering.

And yet.

Imagine Fannie Wall, the striver, lifting, climbing, even gardening twice as well. Imagine Archie working in the yard, back from Berlin, cutting the hedges, weeding, planting flowers. Imagine the Ralphs walking by. Maybe they even had a nice chat with the local celebrity. "Congratulations, son! You must be proud, Mrs. Wall!"

No matter. D-5 would be colored in red on the map, an area officially on the downgrade, according to the Home Owners' Loan Corporation, a division of the Federal Home Loan Bank Board. Such was the power of the *theory* of the city that held sway at the time.

There have been academic debates about the relative importance of HOLC maps in lending decisions and about whether the government initiated redlining or merely sanctioned it. These are important historical facts to sort out. But on the ground, in Oakland, there was one machine built of the same parts. HOLC, FHA, the real estate board. Appraisers trained by Babcock, an FHA guy, might go to the HOLC. Private real estate men made HOLC maps. The boundaries between white institutions were porous, and you could cross them as easily as passing the butter across a table at the Hotel Leamington.

The New Deal was many good things, and some very, very bad ones.

4

The Docks

Among the good things, the New Deal made working people's lives better. The FDR administration's friendliness to organized labor had dramatic consequences for a longtime conflict in the Bay Area between the people who owned ships and the people who worked them.

San Francisco was a port city in the classic sense: cosmopolitan, dynamic, a bit dangerous. The maritime world drew people from everywhere, savory and unsavory alike. A great port required two things: one, a safe harbor, and San Francisco's was North America's best, and two, a workforce to do the heavy lifting of loading and unloading ships. Along the bayside road called the Embarcadero, finger piers reached out into the bay. Each was a world of its own with warehouses, restaurants, and different kinds of workers.

"You walk into Pier Fifteen, there would be gangs working, hundreds of trucks and hundreds of people in and out. Paint gangs, repair people, laundry being brought back," one longshoreman recalled of the current location of the Exploratorium museum. "There would be cargoes from all around the world exposed to your view. Stacks of pepper, spices, flour, grain." Sometimes the cargo was less enticing: most old-timers agree the worst thing to handle was rawhides smeared with manure.

Coffee was one of the big imports. Longshoremen, working in gangs of, say, a dozen people, would descend into the hold of a ship. On the top side, cranes were attached to the sides of the vessels, and riggers would drop a sling down into one of the holds. The belowdeck gang would be split into pairs, each man with a hook that slipped into the burlap of the coffee sack. To move the sacks into the sling, the men would simultaneously insert their hooks and not so much carry but swing it where it was supposed to go. They called the process throwing coffee, and there was a rhythm and a beauty to it, especially with a partner who knew how you worked. Once enough sacks were in the sling, up it went onto the dock, where other workers would get it stored in the warehouse, and then on to the coffee roasters, whose work lent downtown a glorious aroma.

By nature, the work was unsteady, risky, and complicated. Dockworkers had to build the structures in which they stored whatever needed to be shipped. They had to work with the equipment that came on board, which could be of dubious quality or maintenance. They needed to communicate with crews from all over the world, all types of people. And they did it under constant threat of serious injury.

For decades, the longshoremen were like day laborers or gig workers, subject to the "shape-up." In this daily ritual, dockworkers waited and hoped to be hired by the shipping and stevedoring companies. Most didn't know when they'd be hired or how much they'd make. Men who were too focused on safety or preserving their bodies were less likely to be hired. Men whose politics veered too far left might not get work. Some gang bosses expected a kickback from whoever they selected. Once they got hired into a gang, they faced long hours and uncertain conditions. Ethnic groups could be played off one another. Willingness to be exploited was the best way to steady work because there were always more men looking to make some money down at the docks.

On the other hand, a big chunk of longshoremen had what was called "steady" work with one company or another. They were secure relative to the position of the new guys ("prospectors") or other casuals hoping to latch on somewhere. For those with nothing to lose, agitating for change might have been an easy choice. Those who had clawed up to the stability of steady work had more on the line.

Their old union, the Riggers and Stevedores, which was tied up with the International Longshoremen's Association, had been crushed in 1919. Some company unions had arisen, but they did not offer substantial worker protections. Right after FDR was elected, some longshoremen began to organize a paper, *The Waterfront Worker*. They distributed it for one cent. It argued for a rank-and-file union to take the docks.

In June 1933, FDR signed one of the first major New Deal pieces of legislation, the National Industrial Recovery Act. It guaranteed that workers could organize into unions of their own choosing. Galvanized by the legislation, thousands began to sign up with a new ILA local. As the workers organized, a faction led by the charming, irascible Australian Harry Bridges began to push the union in a more radical direction. In October 1933, they struck Matson, one of the three most powerful local shipping companies, after the company fired four union men. They won the showdown and the local union set its sights on bigger changes.

The longshore workers had been up against mobile capital for decades. Because ships could dock in a variety of places, local organizing could not be successful in the long term. The port workers in these different locations had to unite or the network would simply route around them to less troublesome spots. "When one port is on strike, and the ship can move a few miles away and be worked by members of the same union, it's ridiculous," Bridges said.

They wanted an agreement covering all ports. They also wanted the end of the shape-up and the creation of a union-controlled hiring hall that would equitably spread work among the longshoremen. They wanted better pay and fewer hours. Their demands were not new, but their power was.

The conflict intensified through the spring of 1934. In March, the longshoremen organized a strike committee headed by Bridges. President Roosevelt intervened with a telegram asking them to wait for a fact-finding commission's investigation. They did, but after a bum deal emerged, the longshoremen determined to strike beginning May 9. Twelve thousand men walked out up and down the Pacific Coast. In San Francisco, on the very first day of the strike, mounted police officers charged strikers in an ominous sign of what kind of battle the strike would be.

Each side called on its allies. Within days, the other maritime unions had walked out, too, and the Teamsters refused to move cargo off the docks. The workers had shut the port. The employers were not slow to call in the Industrial Association, which represented just about every large local business and the affiliates from national corporations. The police were a constant presence on the waterfront and full-blown skirmishes broke out up and down the coast. On May 28, more mounted police attacked one thousand picketers. There was hand-to-hand combat in the streets. The police were not a neutral force but an arm of the business community.

San Francisco's small Black community was one of the keys to the strike. The shippers were counting on them to act as strikebreakers, the same way they'd been used in other labor disputes across the country and during a 1919 strike in San Francisco. Discrimination and exclusion from the unions made Black people a natural pool of labor for the employers to turn to. At the outbreak of the strike, no more than fifty African Americans worked on the docks where thousands of white people did.

Inspired by a leftist ideology that put class solidarity over racial group, Bridges and his allies recruited Black workers to the union and reached out to the Black community as a whole, famously attending services at a Black church. As Bridges recalled it, he said, "Our union means a new deal for Negroes. Stick with us and we'll stand for your inclusion in the industry." To this day, longshoremen know the lore that Bridges said that if there were two dockworkers left in the Bay, one would be white, and the other Black.

But first, the strike had to be won. In late June, the rank and file rejected a deal reached by the ILA president. On the ground, the situation became volatile. The Industrial Association had taken charge of breaking the strike and it formed a trucking company to "open the port" of San Francisco. Amid thousands of picketers, two trucks emerged from Pier 38, supported by police escorts. The crowd attacked the convoy, throwing rocks and bricks. The police responded with force. The battle lasted four hours, but it was still only a prelude.

After the Fourth of July, four thousand picketers surrounded a freight train. Police beat them back with tear gas, pushing them around the waterfront area and eventually to the union hall. The police opened fire on the crowd and killed two men. The city descended into open street warfare. "Don't think of this as a riot. It was a hundred riots, big and little, first here, now there," the *San Francisco Chronicle*'s Royce Brier wrote. "Don't think of it as one battle, but as a dozen battles." As hundreds were injured in the clashes, Brier reached for the high notes, calling the scene "a Gettysburg in miniature." The National Guard was called in and set up all along the waterfront, machine guns and all.

Four days later, the caskets of the dead strikers were loaded into open trucks. They led a procession of thousands of silent maritime workers down Market Street. The scale and the silence of the procession stunned participants and observers alike. "The sound of thousands of feet echoed up that hollow canyon—nothing else," one marcher remembered. "It was a magnificent sight—those careworn, weary faces determined in their fight for justice thrilled me. I have never seen anything so impressive in all my life."

Even the research director for the longshoremen's archenemy, the Industrial Association, had to admit that it "marked the high tide of united labor action in San Francisco" and that "its dramatic qualities moved the entire community."

The moral drama in the city drew a widening set of players. By mid-July, a general strike had been called in San Francisco. From July 16 to 19, more than 100,000 workers struck. This was not a movie, and the strike resulted not in the capitulation of the employers but rather a monthslong arbitration battle.

But when the National Longshoremen's Board, which had been appointed by Roosevelt, made its decision, the victory for the workers was clear. They'd won the coast-wide contract, the hiring hall, a wage increase, and a thirty-hour workweek. In the fight of their lives, the longshoremen and the maritime workers who joined them had triumphed.

Bridges became head of the San Francisco local, and he kept his promise to the Black community. The International Longshore and Warehouse Union's (ILWU) birth story is not only about resistance to capitalism but about channeling its worst impulses. The shape-up had formed out of the irregular needs of shippers. The day laboring system they created was not unlike what you find outside Home Depots across the nation, or, in a slightly more complicated way, Uber's algorithmic shape-up—the company's fine-grained financial incentives bring more or fewer drivers onto the streets, as its corporate needs require.

The hiring hall, jointly managed by the union and a group of employers called the Pacific Maritime Association, is an entirely different kind of solution. Herb Mills, the longshoreman political scientist, identified a couple major aspects to the hiring hall. It provided stability for longshoremen and protection from exploitation, and helped spread work around more equally. "The hiring hall was indeed the union," Mills concluded.

The ILWU may have been the most progressive union in the country, and Local 10 in the Bay Area became a model of a multiracial solidarity. It wasn't perfect, but at a moment when segregation was an official policy of so many institutions, the ILWU showed a different path was possible, even back then.

5

The Call of California

Frank Tucker was lynched in Crossett, Arkansas, on September 16, 1932. A mob of five hundred tied a noose around his neck, paraded him down the main street, threw the rope over a pipe, and pulled him up to his death. His body was left hanging for forty-five minutes. "The crowd went home to dinner," a news service reported.

Tucker had lived in the town for years with his parents and many sisters. He'd been accused of stealing ten dollars, then while waiting in jail, attacking a sheriff's deputy.

Crossett sits near Louisiana and Mississippi in the southeast corner of the state. It was timber country. Around the turn of the century, the Crossett Lumber Company began buying tens of thousands of acres of land from northern investors. A mill was built. The pine trees that grew in southern Arkansas became lumber, paper, capital, and the town of Crossett. Interlocking companies snapped into place to service the conversion of the natural resources: a small coterie of white families milled the wood, built the company-owned homes, and ran the company-owned store.

People like them logged 20 million of the state's 22 million acres of forest in the early decades of the twentieth century. Ecosystems that had been developing for thousands of years—both those shaped and untouched by the region's indigenous inhabitants—were wiped out in a couple of decades. It was the forestry version of our country's other ecological disasters: the destruction of the Great Plains grasslands and the annihilation of passenger pigeons. The forest of Arkansas became a sacrificial landscape, to adapt the environmental historian Brian Black's term for the early oil fields of Pennsylvania.

Crossett had only a few thousand white residents, so a good chunk of them had been present at the lynching, cheering on the extrajudicial killing of a man convicted of nothing in an American court. A mob is faceless, but Frank Tucker, bound and walking to his death, must have recognized the faces of adults he'd run past as a kid, and peers born, like him, in 1910. Based on the photos from lynchings from around the country, the faces might have been smiling.

Crossett kept growing. Among the newcomers were a little girl, Clemmie Alexander, her aunties Emmer and Clemmie, and her grandma. Otis Whitehead, a laborer at the paper mill, headed the household. Emmer, however, was the motive force. The family lived on Eighth Avenue, in the "colored" section of town. The aunties, women in their early twenties, worked in the chemical plant. Of the men on the block, five worked in the sawmill, five in the paper mill, and two in the plant.

Two young women lived two doors down from the Clemmies and Emmer. Blanch was a maid, Beatrice a hairdresser. The sisters had grown up in the town, their daddy working in the sawmill. Their stories were painful. Beatrice was already a widow at twenty-five. And they'd lost their brother, Frank.

It's not hard to imagine Beatrice might have done the hair of all these women, that they talked about the troubles they'd seen, how that place ran. They must have talked about who was leaving, too, because so many Black people left Arkansas in the following years: fully a third of the Black population of the state left just between 1940 and 1950. An exodus.

War work made California an attractive destination. Airplanes (Southern California) and ships (Northern California) had to be built. Emmer decided to go. She pulled Clemmie Alexander, Margaret's mother, to Richmond, home of the famed Kaiser Shipyards.

They would leave southern racism not for freedom, exactly, but for a northern variant of racism that was bureaucratized and quantified. "I'm a tell you about San Francisco. The white man, he's not taking

advantage of you out in public like they doing down in Birmingham,"
one young man explained in the mid-1960s. "But he killing you with
that pencil and paper, brother."

←---   ---→

Richmond was not an easy place to live during the war. Workers from
all over the country had come to work in the Kaiser Shipyards, as well
as at the gunpowder and chemical plants stretching along the bay.

There were the shantytowns, where people lived "in trailers, tents,
tin houses, cardboard shacks, glass houses, barns, garages, in automobiles, in theaters, or just off in fenced off corners with the stars for a
roof." People survived and worked. It was better to live close to the shipyards, even under terrible conditions, than to commute several hours
each way. The settlement north of the city morphed into an independent Black community, North Richmond. To this day, it remains unin-
corporated. Restaurants and bars sprang up, "rough places that catered
to a hard-drinking, fast-living black clientele," the historian Shirley Ann
Wilson Moore found. For longtime residents, they were dangerous, de-
rided as "juke joints" and "buckets of blood," for the violence and vice
that bloomed within their walls. The music was good, though.

The government built 24,000 units of wartime housing, the largest such project in the country. Most of the buildings were simple.
Two stories. Not nice, just a roof over your head. While some of the
projects were all-white, most were loosely integrated with white and
Black people concentrated in different regions. In one photo, a railroad runs mere feet from the front door of a new building. A Black
woman in a plaid dress and an apron stands on the landing of the
porch, looking back across the acres of housing. Blurred by the speed
of his movement, a man in white pants and a light-colored shirt carries a tiny child. The road is unpaved. There is no landscaping. Yards
are dirt. With the thin walls and the rails running so close, it must
have sounded like the train was going to run you over every time it
went by.

The most desperate situations played out in the schools. There simply were not enough classrooms. Before the war, there had been 6,400 students. By 1944, there were 20,800. Kids were split into groups and assigned a shift, so the classrooms could be used twice. Then a third shift and finally a fourth shift were instituted, so that children were going to school only a quarter of the time. In a photo by Dorothea Lange, for which she asked the kids to raise their hands if they'd been born outside California, perhaps three quarters hold their arms aloft. Margaret's mother, Clemmie Alexander, was part of that swirl of people and activity, if not that exact photo.

The dense war community had its charms. One migrant woman recalled her family's chain of care. Everyone was working, usually working different shifts, so family and friends and boarders were constantly in and out. "Our home was like an open house. Everyone who came went to work in the shipyards and ate some meals at our house," she recalled. "My parents pooled the rations and there was always plenty of everything." The men often gave her and her brother candy, and the women would do up her hair. Women often worked the graveyard shift, then came home to take care of the children.

This was a chance at a new life and people made the best of it. After rural Arkansas, life must have been dazzling, too. In another Lange photo, taken along Richmond's main shopping strip, she captured a woman in front of a store with the sign SERVE YOURSELF. For Black women, who had been confined to domestic labor for hundreds of years, like Archie Williams's mother, this was a world of new opportunity. *Serve yourself*, not others. As one Black California migrant put it, "Hitler was the one that got us out of the white folks' kitchen." The woman looks right at the camera.

In this new terrain, it didn't take long for Margaret's mother to meet her father, Amos Hunter. Hunter was in the Navy, a steward's mate, First Class, like 75,000 other Black men. He was part of the Messmen, which had been the only part of the Navy open to "Negro citizens" before the war, and where nearly half were forced to work during it. In the Jim Crow system, Black people could join the service, but only to serve and feed the white officers. A 1941 government report warned that "the enlistment of Negroes (other than as mess attendants) leads to disruptive and undermining conditions."

Hunter had been a mischievous, fearless kind of kid. His parents had run a garbage business in the hot, dusty town of Anson, Texas, driving a horse and wagon around the streets. That is to say, he came by his cowboy hat honestly. The family was huge: his mother, Lela Mae, was a local legend who raised twenty-three children—nine of her own and fourteen kids she took in. He'd gotten out, though, and made his way west.

Clemmie and Amos married in 1945 and moved in together at 689 Erlandson Street, backed right up on the Southern Pacific railroad line, on the far eastern end of the city's wartime housing. This was where the Hunters would make their first home. They had dreams. If her first child was to be born in falling-down war housing by the train, it would be just a stop on Clemmie's way to something better. She would have other, bigger houses, ones she owned. That was the kind of person she was—moving on up, precise. "I never remember her not being dressed up," Margaret remembered. "She could be going to clean somebody's house. She'd have a crisp outfit. Hair fixed. Lipstick on. Ironed uniform with an apron. And my father was the same way, even with his work clothes. Tears and rips like that? Oh, no, hell no. No, no, no, *no*. Dirty fingernails? Oh no, no, no." Clemmie would raise a Bay Area family, not an Arkansas one, an urban family, not a rural one. Amos would work and work hard, as he did for the rest of his life, in good times and in bad. His children would not want for material things.

From just about the moment Black people arrived in large numbers (or before!), every Bay city had plans to bulldoze the neighborhoods where they lived. So the Hunters, like other Black families, knew that their first home was doomed. The city argued that keeping public war housing would "undermine the value of existing private homes in Richmond," and the state declared all war housing blighted, scheduled for destruction. What refuge anyone had built there hanging on the edge of the continent, freed from official Jim Crow, would be put to an end. The only question was when the bulldozers would arrive.

Margaret beat them there. On December 31, 1946, Amos and Clemmie's firstborn arrived in that war housing, a tiny squall in the baby boom.

6

Spoiled by the Various Mutations of the Air

Why should I fight against the Koreans, Chinese, or any other
Power? Can a Negro truthfully say that he is fighting for the
preservation of freedom and democracy? You cannot give another
that which you, yourself, do not possess. What does the Negro
have at stake in this war?

—PAUL ROBESON SPEECH TO OAKLAND'S NEGRO LABOR COUNCIL,
JUNE 7, 1951 (AS REPORTED BY AN FBI INFORMANT)

In the mid-century, there was a password to run over and plunder Black neighborhoods: *blight*. Blight became a symbol and cause of the country's decline. "For the first time," wrote Mabel Walker in her 1938 tome, *Urban Blight and Slums*, "we are becoming blight-conscious as a people." And once city planners started to look, they found blight everywhere. It was, they thought, the natural consequence of urban growth, a stage of succession.

In the nineteenth century, futures, romances, and pears could be blighted. An early definition rings with poetry: "spoiled by the various mutations of the Air." Blight still lay coiled in its original meaning, a "baleful influence of atmospheric or invisible origin, that suddenly blasts, nips, or destroys plants." This was a word in search of metaphorical use. It never lost its agricultural undertones, but it came to mean "anything which withers hopes or prospects, or checks prosperity."

And white institutions seemed to find blight anywhere Black people moved.

In the public discourse about cities, blight became the intermediate step between healthy neighborhood and slum. It was a progressive disease of urban destruction. Blight killed, even: sociologists and public health officials agreed that poor housing conditions generated civic disorganization and bad health, mental and physical. Worse, this economic disease could not be cured by the tendencies of the market and it threatened whole cities because blighted areas cost more than they paid in taxes.

So it's no surprise that people in West Oakland resisted the label. When an Oakland planner tried to declare parts of the neighborhood blighted, forty homeowners came out to protest at an April 1939 city council meeting, decrying the plans as akin to "Jim Crow," and noting, "Negroes are being discriminated against." That time, though, the designation didn't quite stick.

The city did not stop trying, however. In 1945, California passed the nation's first state redevelopment law to combat . . . *blight*, "a serious and growing menace which is hereby condemned as injurious and inimical to the public health, safety and welfare of the people." The legislators did seem to contemplate the broad new power and tools they were giving their cities. Mayors and city councils could not just willy-nilly designate an area blighted. There had to be a process. A city had to have a planning commission and that body had to have developed a master plan for transportation and land use. Then, those prerequisites met, the city had to produce some kind of report. When the report came back, the redevelopment area could be established and the power of eminent domain unleashed.

This happened all over the Bay Area. In 1954, as Oakland's Black population boomed, city officials succeeded in having big chunks of West Oakland declared blighted and slated for whatever redevelopment the city wanted. That same year, the Earl Warren–led Supreme Court decided *Berman v. Parker*, a crucial case about the power of cities to control their land.

The court ruled that local governments could take private property that was within an area that had been declared "blighted" and do whatever they pleased with it. Building on a landmark zoning case, *Village of Euclid v. Ambler Realty Company*, the justices gave cities carte blanche in the name of protecting the property and prosperity of their citizens.

"Miserable and disreputable housing conditions may do more than spread disease and crime and immorality. They may also suffocate the spirit by reducing the people who live there to the status of cattle," wrote Justice William Douglas. "They may indeed make living an almost insufferable burden. They may also be an ugly sore, a blight on the community which robs it of charm, which makes it a place from which men turn. The misery of housing may despoil a community as an open sewer may ruin a river." In practice, this meant that bulldozing a blighted area was, *in and of itself,* a public good, based on the sound advice of planners.

The Supreme Court decided *Berman* just months after the famous school integration case *Brown v. Board of Education*. While the *Brown* case has been celebrated as a crowning racial achievement of the mid-century, *Berman* also had an outsized impact, but in the opposite direction. Property owned by, or even near, Black people was suddenly "less worthy of the full bundle of rights" that white Americans assumed, wrote the legal scholar Wendell Pritchett.

The discourse of blight was the perfect tool for disappearing race into the scientism of the planning process at precisely the time when the Supreme Court was beginning to frown on obvious and outright discrimination by race. Stepping back historically, it seems like a racist masterstroke. By maintaining the lines of segregation during and after the war, forcing Black migrants into just a few small areas, the white urban power brokers helped *create* the problems that they could then name *blight*. Real estate and downtown interests had a proxy name for Black neighborhoods, one they could use to wield the increasingly powerful hand of government without threatening their own property rights.

In Oakland, the culmination of this process was a report called the "Oakland Residential Area Analysis." The analysis assigned points based on ten "indicators of residential quality," most of which had been borrowed from San Francisco or the "Hygienic Housing" research of the American Public Health Association. These factors were combined into an index called the "penalty score." This idea was borrowed from the "Hygienic Housing" research, but it also had an antecedent in the FHA's scoring system.

To be fair, there were very serious problems with West Oakland housing. Largely white landlords had made a killing piling Black people into substandard housing without upgrading or repairing the buildings. But Oakland did not try to target those landlords' holdings. They took whatever data was at hand and applied a byzantine computation, adding up a myriad of weighted calculations. Then the census tracts' penalty scores were grouped into quartiles, and shaded on a map.

The *Oakland Tribune* put its key urban affairs writer, Bill Stokes, on a story about the analysis. He crafted a dark introduction. "A total of 1,140 acres of Oakland's residential areas are in a slum or badly blighted condition," Stokes wrote. "Another 1,406 acres are almost as bad." Together, those figures added up to the headline figure: "24 Pct. of Homes in City Blighted." This was a crisis. A whole quarter of the city!

Stokes was a city insider, a guy who would become the head of BART, and his interpretation of the numbers almost certainly reflected how elites saw things: 97.5 percent of Oakland's Black people resided in what the *Tribune* called the blighted area. So: 46,390 Black people were in it and 1,192 were not. While white people also resided in some of the districts declared blighted, only one census tract in the least-penalized quartile had more than 1 percent Black residents (2 percent of its residents were Black).

As I worked through these documents, rerunning the calculations from the old statistics, trying to make sense of what the city was doing, I came to my own conclusion about blight. It was a real thing. There were areas of the city that were like a sewer, in part because they were located near the sewage plant or they didn't have indoor plumbing or they were polluted by industry. "Blight," as calculated by the city, was an index of environmental racism.

Look at the factors that often went into calculating whether a neighborhood was blighted: rising density, increasing inhabitants per home, pedestrian deaths, juvenile contact with police, disease prevalence, deteriorating housing stock, mixed industrial and residential uses, falling land prices, rising city costs. Racist structures shaped every single one of these conditions. Blight was a way of measuring what had been done to the people of a place.

Black areas were dense because there were many new Black residents in the Bay, but few neighborhoods in which these newcomers were able to live. Black housing was often older because those were the regions where the real estate industry showed Black people houses. Decades of exclusion from unions and discrimination held Black wages down, so most could afford only lower rents. Black people could not access capital at the same levels as white people, if at all, so they were unlikely to be able to afford the newest homes, even if they could slide past the official and unofficial racial restrictions. Residents don't ask to live next door to factories or to be hit by cars or for the police to arrest their children.

After the residential analysis report came out, Oakland declared 250 blocks of West Oakland to be part of its redevelopment program. The first patch to be condemned to destruction was on the south side of Seventh Street, just across the street from the Slim Jenkins Supper Club. The site was to be the home of a new post office distribution center.

As the inner city fell apart, prices for housing *did not* fall, as the segregated market allowed landlords to continue to charge Black people high rents. But the more Black people that moved into the old city, the higher property values rose in the new developments that the FHA backed in the 'burbs. The more marginalized racial and ethnic groups that moved into the inner city, the greater the "rise in the values of land in the outer areas of the city," a Federal Housing Administration economist noted.

Or as Ms. Margaret once put it: "You got to look at how devastation for some gave opportunity to others. One was down, one was up."

◄┅┅   ┅┅►

In the mid-1950s, Oakland was hardly the port city of the romantic imagination. That was San Francisco, of course. In Oakland, the real action was at the Oakland Army Base, which could accommodate sixteen ships at a time. Commercial ships came and went, but they just stopped by on their way out of the bay to pick up produce from the great Central Valley.

The city's shoreline had been dominated by shallow marshlands. On the south side, a freshwater creek had been turned into a brackish estuary by dredging, creating the island of Alameda. To accommodate large military ship traffic, deep channels, thirty feet or deeper, had been grooved into the otherwise shallow mud of the Oakland shore. Two harbors—Outer for the Army and Middle for the Navy—had been created by reshaping the bay.

The water was as much an impediment to commerce as a boost. Multiple plans contemplated filling substantially the whole bay to accommodate more development. The "bay" would have become something closer to the Los Angeles River, and it wasn't totally impossible to imagine. Outside the shipping channels, the bay was often less than ten feet deep, sometimes less than five.

The bay was also an environmental disaster. Dozens of sewer outfalls poured their untreated waste into the water surrounding Oakland. The water reeked. It was almost a relief when the winds blew the smell of Heinz ketchup from a West Berkeley factory down the shoreline. The entire area near the shore was zoned for industrial use, including more than 550 acres filled with housing in West Oakland, not to mention the entire Seventh Street strip, the center of Black life.

On a financial level, the Port of Oakland's business was sluggish, and not forecast to grow much for years. But the city had big dreams. The planning department issued a beefy report in 1952. Above all, and with admirable foresight, Oakland officials realized that the port's future depended on the economic development of Asia. "The industrialization of China and the rest of the Orient is late in arriving. Because of its magnitude it will inevitably produce fundamental modifications in the world's economic and political structure," the report declares. "No one can say exactly what the impact of the industrial revolution in Asia will be on the economy of the Pacific Coast; the potentialities are beyond imagination."

That potential was, in fact, too great for San Francisco. The only way that the levels of trade between a rising Asia and a dominant America could reach their potential heights would be the development of Oakland as a port. There just wasn't space in the big, dense, old city. It would be the East Bay tidelands generally and Oakland specifically that would physically tie Californian businesses to their Asian counterparts. "These lands are near existing transcontinental rail lines, have direct access to the large East Bay labor market, and from the regional point of view are ideally suited for industrial locations," the report argued, "assuming problems of air and water pollution are overcome."

The industrial development of Asia, however, was not some passive process for the United States. In a pair of top-secret 1949 memos, NSC 48 and 48/2, the State Department laid out "The Position of the United States with Respect to Asia." The U.S. would maintain "liberal" trade policies that encouraged "stimulation of imports from Asia." American private capital would be encouraged to flow to friendly countries there, both directly and through international institutions. American planners thought encouraging "capital movements, trade, and other economic relations" was paramount. Specifically, Japan had to be built up as a base of anti-Communist stability, and "to demonstrate the advantages of close association" with the United States. At the same time, the U.S. government would support the aspirations of the people of Asia only "in such a way as to satisfy the fundamental demands of the nationalist movement while at the same time minimizing the strain on the colonial powers who are our Western allies." It was no accident that first Japan and then many Asian countries became powerful manufacturing centers that relied on exports to the United States.

The desire to maintain control over Asia led to the U.S. involvement in the Korean War, an awful and vicious conflict that the United

States could not win. The war killed a million civilians in North Korea, perhaps 15 percent of the population. A million people! American bombs destroyed more than 50 percent of the vast majority of North Korea's large cities. Having perfected the techniques for firebombing cities in World War II, the Air Force showed no mercy or sense of proportionality in laying waste to the country. "We killed civilians, friendly civilians, and bombed their homes; fired whole villages with the occupants—women and children and ten times as many hidden Communist soldiers—under showers of napalm," said an American *official history*, "and the pilots came back to their ships stinking of vomit twisted from their vitals by the shock of what they had to do."

All the horrors of napalm, which would come to be associated with Vietnam, were visited upon Koreans first: charred skin, writhing children, indiscriminate destruction. North Koreans were forced underground. International observers found whole cities where no one—no one!—had a building for a home. A family might share a small mud hole dug into the ground. A family might be some older children caring for their younger siblings, their parents dead.

Douglas MacArthur, the general who had first been in charge, told a Senate committee that there was no comparison to what we did in Korea. "I have seen, I guess, as much blood and disaster as any living man, and it just curdled my stomach the last time I was there," he said. "After I looked at the wreckage and those thousands of women and children and everything, I vomited."

How did the United States end up prosecuting this horrible war? At the end of World War II, Soviet and American forces divided the country, somewhat arbitrarily and to the benefit of the United States, at the 38th parallel. Korea, with a thousand years of distinct heritage as a proto-national people and place, had been under Japanese imperial occupation since 1910. Freedom fighters had tried to liberate the country with limited success. They were still considered heroes, while those who collaborated with the occupying Japanese were seen as something like the French collaborators with the Nazis. The problem, from the American government's perspective, was that the anti-imperialists also tended to be Communists, in particular, Kim Il Sung. The conservative politician the United States ended up backing, Syngman Rhee, had been in exile for decades.

Rhee exercised power as a dictator. For the years following the war until the hostilities took off in the middle of 1950, his government, working with U.S. backing, killed suspected Communists and put down revolts and uprisings. Political violence killed 100,000 South Koreans before the war, as the historian Bruce Cumings pointed out. Tens of thousands died on the island of Cheju alone, as fascist youth corps, police, and the military deployed vicious anti-insurgency tactics. "It was on this hauntingly beautiful island," Cumings wrote, "that the postwar world first witnessed American culpability for unrestrained violence against indigenous peoples fighting for self-determination and social justice."

As the two sides fought to the bloody stalemate that has given us North Korea and South Korea, the Korean War was a boon for American companies. Military spending quadrupled; amazingly, it reached 15 percent of American GDP. Today, with the much larger American economy, that would be *$3 trillion*, more than four times our actual, highest-in-the-world military spending. This was nation-shaping-level money.

An outsized chunk of those dollars landed right in the Bay Area. Much of the cargo for the war flowed through Oakland, keeping the Army base humming at a greater average output than during World War II and setting the stage for Oakland to become the test case for transpacific container operations during Vietnam. In Palo Alto, Stanford had been refashioning itself from sleepy suburban also-ran into what the historian Margaret O'Mara calls the "perfect laboratory for the military-industrial complex." The university's dean of engineering, Fred Terman, had been a crucial science administrator in World War II electronics development, and his expertise and connections served the university well. He focused research on microwave tubes, precursors to the transistor, because he knew how useful they were to the military. Meanwhile, he encouraged companies to settle around Stanford, where they could commercialize all the new engineering knowledge. Local spinouts such as Varian Associates, Litton Industries, and, of course, Hewlett Packard would grow alongside the university, watered by military dollars. Vast new divisions of national companies put down local roots, too: Lockheed, IBM, GE, Kodak. The business of the valley would become radars, missiles, and bombs.

Terman and the scores of other men who'd gone through World War II retained their belief in the importance of planning and public-private collaborations. This was a model for how research should be done. Stanford was *supposed* to be in business with the war industries. That kind of collaboration had defeated Hitler. And they imagined they'd do it again with their new enemy, the Communists.

They had counterparts throughout the U.S. government. The West Coast became the nerve center for the American system in the Pacific. Electronics manufacturing would begin shifting to Asia, a process that continued for the next half century. Silicon Valley was in a perfect position to take advantage of this new labor force, and as electronics grew ever more important, the technology companies of the Bay Area would become the tie between federal government R&D, military needs, foreign policy action, and consumer products. Electronics would help America win the war *and* win the peace.

High-technology companies would create and use new logistics systems to create an unprecedentedly powerful economic loop that bound growing Asian countries to the corporate and physical infrastructure of the West Coast. The rise of the technology industry is as much a story about this incredible transnational exchange as it is one about delightful nerds at the Homebrew Computer Club.

It should come as no surprise, then, that the Korean War, this brutal war, was good for the Bay Area. One was down, one was up. Our economic geography is, to a great extent, the long-term consequence of the U.S. government working with industry to shape global affairs. The factories sprouting up in Asia were as much a part of the plan as the suburbs flourishing around Stanford. Even before silicon arrived in the valley, this was supposed to happen.

7

One Was Down, One Was Up

The newly built suburbs were a boon for white people in the Bay Area. In Northern California, researchers from the U.S. Commission on Civil Rights found that fewer than a hundred non-whites were able to buy into the 350,000 suburban tract homes that were built in the years leading up to 1959. That's 0.029 percent. The FHA played a determinative role in the building of those subdivisions, and scholars such as Richard Rothstein have shown how important racial homogeneity was to their willingness to back a project.

Meanwhile, the Bay Area's local governments and residents did their best to enforce segregation, too. For example, the Richmond City Council laid out its redevelopment agenda in 1950. It proposed bulldozing thousands of units of war housing, including where Amos and Clemmie lived, while providing "protection" to single-family home areas, which were nearly exclusively white. "The long-established ideals of America are represented in these neighborhoods of small, pleasant, single-family homes," the 1950 redevelopment report held. "These home-makers, who ask for little, deserve every encouragement and every bit of aid that can be offered by their government." And, of course, one part of that aid was a pledge to "maintain the harmonious residential character of neighborhoods." *Harmonious* was a code word for "racially homogeneous," with its roots in the Chicago real estate world. Homer Hoyt, who held a University of Chicago PhD and worked on appraisal methodology for the FHA, used *harmonious* to define the kind of neighborhood that was the pinnacle of the American dream. "A neighborhood composed of new houses in the latest modern style, all owned by young married couples with children, is at its apex," he wrote for the FHA. "At this period of its vigorous youth, the neighborhood has the vitality to fight off the disease of blight. The owners will strenuously resist the encroachment of inharmonious forces because of their pride in their homes and their desire to maintain a favorable environment for their children."

If blight were merely the physical deterioration of property or the impingement of polluting facilities, Hoyt's description might sound morally neutral, even virtuous. But in the context of the mid-century housing market, the reality was much uglier. What did it actually mean to "resist the encroachment of inharmonious forces"?

A Navy man like Amos, Wilbur Gary, his wife, Borece, and their seven children were living in the Harbor Gate projects, right on the eastern edge of the war housing developments, practically neighbors to the Hunters. Knowing that the housing had been slated for demolition, and that other projects had already been destroyed, the Gary family found a white Navy man to sell them a home in Richmond's FHA-backed, all-white Rollingwood development in 1952.

When word got out, a burning cross was placed on the lawn, Klan style. When organizers with the Congress of Racial Equality, which was a Communist front group, showed up at the house, they found a white crowd chanting, "Out, niggers!" A woman in the crowd shouted, "My property would go down two thousand, five hundred dollars if the niggers moved in." This is what a neighborhood at its apex was *expected* by the federal government to do.

The situation was volatile and truly dangerous. Anti-Communist FBI informants reported that the Congress of Racial Equality had been "organizing patrols to guard the [Gary] family on a twenty-four-hour basis." Eventually, the NAACP and local white families helped disperse the vigilantes. The Garys settled in and raised their children. But that's what the all-white suburbs were like in those days. It took hard work, guts, and luck to beat the cordon holding Black people in their place.

After Margaret was born, her family moved to public housing in San Francisco's Hunters Point, right near the shipyard that was the center of the area's economic life. That first house on Erlandson was demolished soon afterward.

In Hunters Point, their neighbors were nearly all southern people with California kids. Texas, Arkansas, Florida, Virginia, Missouri, Mississippi, Georgia, Washington, DC. The great reshuffling of the Great Migration. To this day, Margaret has the vestiges of a southern accent. The neighbors worked at a grab bag of working-class jobs: mail handlers and laborers, military men and machinists, janitors, clerks, and seamen. They had white neighbors, too.

Her parents were busy: Margaret had two little sisters by 1950. Her dad worked as a mail handler at the U.S. Post Office. Margaret remembered mostly seeing the back of his head as he headed out to work. "He was always working," Margaret said. "He always had a job and a side hustle." Her mother kept house at 47 East Point Road, unit 916.

They bounced around San Francisco between Black neighborhoods, looking for something more permanent. By the late 1950s, they had saved enough money to think about buying a house of their own. Amos's veteran status helped. He used his GI Bill benefits and went through a Navy buddy to help him find a house for his growing family. It ended up being in Ingleside, 330 De Long Street, way at the south end of San Francisco, nearly at the border with Daly City. "We were like the first Black family on that block," Margaret said. "Then, you know, you saw the white flight."

The house was nothing special. Just another two-story row house slapdash attached to all the others on the block. It'd been expanded at some point, even if that wasn't in the official plans, so it was big. The garage was on the ground floor. A flight of steps ran up to the front door.

"That house at 330 De Long Street. That was the hangout house. Every holiday. Every birthday. Everything," Margaret remembered. "In the backyard, in that den, or in that garage, we had something going on. It was always full of family and friends. Always full of somebody."

For a young Margaret, life in Ingleside was great. Many new Black residents were arriving in the neighborhood. They were mostly middle-class, making a life out there in the fog, away from the grit of Hunters Point and the action of the Fillmore. She and her friends all went to the same set of schools: Sheridan for elementary, James Denman for middle school, and then on to Lincoln for high school.

It was nice for Margaret's parents, too. If one of the girls was getting into trouble, they could rest assured that somebody's mama was going to make sure she got set right and marched home. Her father could take a hot shower after a day at work and get started on dinner, something warm to beat back the cold wind out there. The first meal Margaret remembered cooking with him was a feast, drawing on his culinary training in the military. Roasted duck stuffed with apple slices, ginger caramel carrots, wild rice, and asparagus. But he had range, too. He liked to make homemade pizza, try out Chinese dishes. Always lots of vegetables. He'd come a long way from that garbage cart in Anson, Texas.

Here was a place of sanctuary, the end of a long climb.

Many other Black people in the Bay Area found a place to hold on to in those years between the war and the early 1960s. It wasn't always as cozy as the Hunters' house in Ingleside, but they were places of respite nonetheless, where the community could cohere after the migrations of the previous decades. Homes, bars, little joints run out of someone's house, fancy restaurants, BBQ spots. All of it added up to more than the sum of the parts. It was community and economic infrastructure.

The city of Oakland did not have the eyes nor the inclination to see the value of what the Black community had built in West Oakland. The big shoreline development report did not invest much time or energy in understanding those who would be harmed to let Oakland receive the rewards of Asian growth. In its analysis, all of Seventh Street's vibrancy, attested to by residents and open-minded visitors, was reduced to a few lines noting the "small retail stores and wholesale outlets." The longest-running Black neighborhood in the Bay became merely "dwellings in poor condition with stores intermixed." Without reservation, the whole area to the west of what would shortly become the Cypress Structure was "completely unsuited for residential use," even if many thousands of people were *already living there.*

What had Black people managed to build in the area that was

"completely unsuited for residential use"? In the days before the Great Migration brought tens of thousands of Black people to the city, the west of West Oakland was a deeply mixed community. Mexicans whose families had come north, Chinese coolies, Portuguese railroad workers, German shopkeepers, Irish laborers, and whoever else found themselves down by the tracks. Stare at a picture of a Prescott kindergarten class from the early twentieth century and you might think you'd fast-forwarded to a future of racial harmony.

By the mid-century, Seventh Street was the center of Black life in the East Bay. There were dozens of restaurants and clubs, including the fanciest, the Slim Jenkins Supper Club, and Esther's Orbit Room, which is still standing.

A new species of blues evolved in the clubs and in the recording studio Bob Geddins built on the east end of the strip. In the 1950s, on just the block between Willow and Campbell, you'd have found the Kit Kat Club, Club 4, Bob's Pool, Doggie's, John Henry's, and the Beulah Club. Slim Jenkins was just one block down.

The proprietors of the clubs had built themselves from nothing, and were experts in the narrative arts. Harold "Slim" Jenkins ran Oakland's most famous club, at Seventh and Wood. He was a tall man—erect and fastidious. He cultivated an upscale, integrated clientele. On his business card, "Slim Jenkins" curved through an elegant, unusual script. "Good Food ★ Better Entertainment / Specializing in Southern Fried Chicken," read the top. You'd even find the clubwomen there, tables of them gathered under a sparkling sign that read DAISY BRIDGE CLUB.

These spots were also where the formal and informal economies met. Rumors that the local gambling (and Prohibition booze–running) kingpin Charles "Raincoat" Jones actually financed the place never quite went away. In fact, Jones actually owned the establishment right next door to the supper club, the Sportland Recreation Hall. It was raided many times during the 1950s, much to the dismay of Jones, who told the *Oakland Tribune* that the young police just did not "know the score." Police maintained there was a dice game in the large back room, and there almost certainly was. Not that the possibility of connection between Jenkins and the underworld would have necessarily been a drawback for patrons—the hint of danger is part of what makes nightlife anywhere, but especially back then on Seventh Street, fun. Anything could happen.

The connections between and around the bars were not only illicit. They were key nodes of the systems that developed under the segregated economy. Esther Mabry began her career as a Jenkins waitress, dressed in a smart shirt and long skirt, before running her own bars in West Oakland. In an oral history late in her life, she recalled that the Black cabdrivers had created an organization for themselves, Associated Cab, that met near her first place. The railroad men had the Pullman porters' union. The cooks and waiters had theirs, the seamen, the longshoremen, too. All of them would meet at Esther's. With so many people flowing into town and so little official government infrastructure to help, Esther and people like her were an interface into the city for newcomers, a crucial part of helping people adapt to the new place. "I would get jobs for people," she remembered. "I knew just about everybody."

That economic cohesiveness helped connect local people to political power, too. FDR was said to have patronized Slim Jenkins. The legendary California assemblyman Byron Rumford and Governor Pat Brown would come have dinner at Mabry's place. The Democratic Party would hold meetings there, hoping to oppose William Knowland's Republican machine, which preferred Jenkins Supper Club.

At night, the streets filled with people, and the culture tilted toward strolling, popping from one place to the next. Southern food places cooked up what people were missing. They had fun. They got drunk, smiles wide, faces shiny, eyes bright. The mythology and reality of those times are hard to separate. Oral histories mostly recall a smear of fun and hijinks: different musical styles and class dynamics and regional foodways all get melted together into one mass of memory about the old days. It was what Black communities had and were and imagined themselves to be at that time. It's no crime to remember those days in the fondest terms without any grit or gristle.

Margaret's parents loved to go out, and you can understand why. If you were from Anson, Texas, or rural Arkansas, what greater reward was there for leaving all that you'd ever known than being able to walk down the street, go grab a drink (maybe from the bartender Ted Jernigan at Slim Jenkins), hear some music (Jimmy McCracklin, say), dance till the sweat bled right through your pressed shirt, and then wobble out into the street to Jumping George Barbecue for ribs? California wasn't everything that anyone dreamed, but no one could take a night like that away from you.

The memories last. But as it turned out, the whole neighborhood could be swept away.

8

Killing You with That Pencil and Paper

On August 15, 1960, Abdo S. Allen arrived at Third and Peralta in West Oakland. Six two-story homes stood at the corner, adjacent to the Southern Pacific Rail Yards, and a few blocks from the heart of Seventh Street. From a truck, he unloaded a Sherman World War II tank, put on a helmet, and climbed inside. He maneuvered the tank directly in front of the first house as a crowd watched. Then he plowed his way through the ground floor. The tank emerged, covered in dust and debris. The first house took ten minutes. All six, just an hour and a half.

The town's white elites—mayor included—had gone to the ghetto to watch the show, which served as a groundbreaking for their sluggish urban renewal plans. Black children in the neighborhood gathered, too, watching the destruction. Paul Cobb, a future Oakland civil rights leader, lived in one of those homes. Huey Newton was a boy in the neighborhood. None of their reactions was recorded. Newspaper editors across the country jokingly headlined stories about the destruction, "The Battle for Oakland." They noted the tank cost Allen $2,000 and he got paid $64,000 for the job.

The demolition was the culmination of six years of quantifying the case for leveling the neighborhood, two years of negotiations between the city and the post office, conducted in secret, and another year of official planning after the announcement of the project the previous August. The post office promised a huge, new high-tech postal distribution center modeled on a fourteen-acre site in Rhode Island that

Eisenhower's postmaster, Arthur Summerfield, had said would be "the first automatic post office." Oakland's version would also be a "revolutionary post office and working laboratory," filled with new electronic devices. Gateway, as it was known, would be the center of a new vision to bring automation to the USPS.

For Oakland city officials, this was a way to jump-start redevelopment. Right in the heart of what they saw as the Black ghetto, they could wipe out twelve blocks in one go, funded by the federal government. Because the area had already been declared blighted, there was no real problem obtaining the parcels of land. Some they got from the railroads; 126 others they condemned in federal court on the day that they announced the project, describing the area as "mostly substandard housing, with a few junk yards and some commercial buildings."

Some people, however, were looking past the smashing of the homes with the tank. People *lived* in those homes. Some had their rental homes sold out from under them. The Oakland Redevelopment Agency developed what the *Tribune* called a "crash program" to help people find housing. They assigned Dr. Helen Amerman, the agency's relocation officer, to the task. She'd earned a doctorate in race relations from the University of Chicago and had helped author a major report on race and housing in the country, as well as serving as an officer for the Council for Civic Unity in San Francisco. It was actually Amerman's second relocation job. During World War II, she'd worked inside the Minidoka War Relocation Authority, counseling incarcerated Japanese Americans on the outskirts of Twin Falls, Idaho.

The post office refugees were folded into the planning for the adjacent Acorn urban renewal project. City officials commissioned a study of the housing needs of the residents of the areas slated to be bulldozed for the two projects. When the city finished tallying up the 1,650 families that would need to move, their analysis found that 502 families would need low-income public housing. This was *some coincidence*, as the total nearly perfectly matched the number of public housing slots that the Oakland City Council had voted to build in 1950. Thus, the city could tell the federal authorities that it had a workable relocation plan, which it was obligated by law to create.

←---   ---→

The plan for the automated post office in Oakland didn't come out of nowhere. It was supposed to solve what we would now call the major logistics problem of the time. Paper had become the lifeblood of the huge institutions that dominated the mid-century. Corporations and governments lived on paper. It moved the mechanisms of business and government. Eight million people in the mid-1950s were paper pushers of one kind or another. The residual shudder tucked inside the word *paperwork* still conveys the horrors of the bureaucratic sublime. "Sales slips, purchase orders, payrolls, insurance accounts, auto registrations, income and social-security tax-returns, railroad and airplane and bus reservations, bank checks and statements—these are the sort of thing that make up the flood," one writer listed.

This is the world of the Terry Gilliam movie *Brazil,* in which ductwork for moving paper has become architecture. In this predigital world, the means for transmitting information, in the form of paper, had become ridiculous. The flood of paper was a sign of a deeper problem: moving things took too long, when the *idea* of those things could move so quickly via telephone. This reality began to shape West Oakland. Goods movement had to happen somewhere, and it wasn't going to be in downtown San Francisco.

The Post Office Department, as it was known, faced a surreal version of the paper crisis. The incoming postmaster in 1953, Summerfield, discovered that if he wanted to know the results of operations from the previous month, it would take nearly a year and a half for the organization to compute those figures. Half a million employees handled 49 billion letters in 1953, along with a billion parcels. Simple projections of increased mail volume seemed to show that in the future, as one writer joked, "half the citizens of the United States would be delivering mail to the other half."

Summerfield instituted a raft of reforms, setting up a more regional structure that was capable of creating "statistical by-product data for management analysis, planning, control, and decision-making." This, in turn, set the organization up to take advantage of the emerging new technology of "computers."

The automation of mail handling would prove to be far more difficult. Until Summerfield's tenure, all the logistics of moving mail from intake to delivery had been manual. There was a reason for this. The address that sat on every piece of mail was a complex, legible-only-to-humans object. To make automation work, the address had to be converted from pen marks on an envelope into correct data for automated processing.

With the technology of the mid-1950s, it was *juuuust* imaginable that this would be possible. Character recognition with what were referred to as "electronic eyes" *seemed* promising. And it was in this spirit that Summerfield began to attack the problem of mail sorting. He dreamed up a new kind of distribution center, first to be deployed in Providence, and shortly thereafter in Oakland. These were to be "revolutionary post offices . . . built to accommodate every mechanical and electronic device practicable as replacement for human legs and hands." As Summerfield imagined it, machines would be put to work at every step of the way, commands given out by a "control tower" not unlike "the control of a great airport."

But it did not work. The Providence project was a dismal failure. Like many other early artificial intelligence efforts, people—even experts—wildly overestimated the capability of the available machines. People, as it turned out, could do amazing things that machines would not duplicate for a long, long time.

Construction in Oakland was put on hold. The twelve blocks that had been bulldozed along the Black community's main commercial thoroughfare would remain empty, a scar on the landscape.

Through the years, Amos Hunter kept Margaret's family going financially. Along with the various side hustles, he worked at the post office in San Francisco and also as a mail handler at the Oakland Army Base, right there in West Oakland. The Oakland Army Base was a town-sized facility back then, employing perhaps five thousand people, including several thousand civilians and waves of longshoremen. It was a huge part of Oakland's economic base, and it was the key port for the military in the Pacific theater. All kinds of stuff shipped out through those docks: weapons, food, tent pegs, radios, shoelaces, shaving cream, letters for troops.

As far back as the mid-1950s, all the branches of the military were desperate to understand what they actually had in their warehouses. There were, for example, 1,254,700 items in the Navy Supply System, and the total value of the goods it warehoused was $13.7 billion, all managed on paper. As the capabilities of computers began to be more widely known, it was extremely tempting to imagine that all those items could be stored in a computer, which would allow for the kind of analysis Summerfield wanted at the postal service.

This would become the field of *logistics*, right alongside "operations research" and "numerical analysis," a branch of mathematics closely associated with modeling.

"A superficial view produces the idea that logistics is military economics," Captain Charles Stein wrote. "It is really much more than that, embracing as it does a complete range of down-to-earth operations with hardware and hard-handed operators to highly subjective planning and forecasting functions."

Nearly all the early computers of the mid-century can trace their lineage back to military needs and funders. ENIAC, the first electronic general-purpose digital computer, was designed at the University of Pennsylvania by John Mauchly and J. Presper Eckert. Mauchly and Eckert then spun out a company. The initial order for their new machine—the first real commercial computer—specified three units: one for the Census Bureau (funded by the Army) and two for the Air Force Materiel Command, the area of the force that dealt with logistics.

The odd early machines also included the Logistics Computer, which was operated out of George Washington University's Logistics Research Project. The Office of Naval Research ran simulated war games using the Logistics Computer. The Air Force followed suit through the RAND Corporation, setting up a series of increasingly complex logistics simulations.

Amid the electronic buzz, at least for the Navy, there was still the problem of what lay behind the data, as Stein indicates, the "hard-handed operators" part of things. The "cargo handling" research was assigned to the engineering department at UCLA, under Russell O'Neill.

While O'Neill admitted there was a lot of "know-how" about cargo handling, it had not been subjected to the rigors of professional analysis. There was "little formulated information" on the topic. Longshoremen did most of the work and they were not the kind of people whom engineers consulted. The gangs that worked the ships were like teams with different kinds of expertise and skill embodied within the men.

But O'Neill could not help but see cargo handling as pathetically primitive. "Cargo handling has remained essentially the same since the days of the wooden ships," he wrote in his "Engineering Analysis." "Cargo is still taken piece by piece and hoisted with a hook up and over the side of a ship and into the hold."

He knew that, in the details, this was "extremely complex," and set about "to order these complexities and find the relations between them."

To longshoremen, the waterfront had already seen encroaching change. Most important, bulk commodities had experienced ever more mechanization. For example, the Matson Navigation Company had long shipped raw sugar from Hawaii in huge sacks to a processing facility in the Bay, at Crockett. It was remarkably labor intensive, requiring 6,650 man-hours to load 10,000 tons of sugar. A new bulk conveyor system cut the labor need down to 1,000 man-hours. There went hundreds of days of work for dockworkers per ship. Similar things were happening with grain, ore, paper, even bananas, to say nothing of oil, which was moved in tankers that were rapidly scaling up.

Even handling break-bulk cargo, as it was known, was not precisely the same as it had always been. Forklifts had been in use since the 1930s. They worked in concert with another crucial innovation: the pallet. Together, they made longshoremen faster and preserved their bodies.

Over time, the goods that were being worked changed, too. More and more often, people were shipping finished products, not raw materials. These needed protection and to be kept together. This resulted in what was called the "unit load," stored in a pallet or in different types of boxes. "The containers that are used vary a great deal from one situation to another," O'Neill noted.

In the early 1950s, the problem for shippers—military and civilian— was obvious. Cargo ships took too long to load and unload. O'Neill began to formalize the system, however, creating equations for modeling how longshoremen handled peaches, for example.

As he went, the whole picture of cargo handling became more and more conceptual. The work of the men, the strain of time spent working deep in the bowels of a ship, the feel of different cargoes all fell away. And in return, he was able to note areas where efficiencies might be achieved. He sought optimization. At O'Neill's most abstract, he even created a hypothetical electric circuit as an analog to the work of longshoremen unloading a ship.

Nothing could have been more removed from the mess and noise and bodies of the docks. Which was, in fact, the point. The variability of real-world cargo introduced complexity that the circuit couldn't actually capture.

But what if every piece of cargo was the same? A box on a diagram would be a box in the world. If everything that went onto the ship was a standard size and configuration, many of the variables that complicated the analysis of worker productivity, facility efficiency, and, ultimately, cost would disappear.

Was it difficult to imagine such a system?

The answer was, more or less, no.

Graham Brush created a version of it in Seatrain Lines in the late 1920s. Running from New Orleans to Havana, Brush's ships had four railed decks. A gantry crane on the pier would pick up railroad cars and drop them into place. Another crane on the other side of the journey would reverse the operation. Seatrain Lines could turn a ship around in ten hours, a fraction of the six days a comparable regular cargo ship might take.

To industry insiders, the potential of the system was clear. Seatrain augured "a complete scheme of transportation, including rail, truck, and ship, providing door-to-door service." Alaska Steamship began to offer just such a service in the late 1940s on the Seattle–Alaska route, carrying truck trailers, which it passed to other companies for delivery over the highways.

Implementing these systems, however, was not easy. There was no standard container that would fit on a truck, ship, and railroad. There were no ships that could carry large numbers of containers. There were also no piers that could deal with that kind of operation. No one knew if a container-bearing ship should have its own crane or rely on dockside facilities. And within the United States, the Interstate Commerce Commission mostly ignored, if not outright frowned on, "intermodal" cargo. Truckers trucked. Shippers shipped. Railroads railroaded.

The breakthrough that would unlock the Pacific Circuit was the creation of a network of interchangeable packets of stuff. All the boxes would work together.

9

Right-of-Way Men

The kind of research and planning that went into the military's cargo system typified mid-century urban life. This was the golden age of top-down planning. Men who had been trained in operations research and logistics during the war wanted to apply the same kind of thinking to optimize and tame the unruly city. That is to say, the elites of a region could be counted on to issue normative reports on the direction of the metropolis. And central to those plans in San Francisco was a new transportation system.

Long before it was functioning, BART was an idea about how a transit system could be used to shape the region. San Francisco's downtown would dominate the bay, becoming, as one skeptical reporter put it, "a second Manhattan" that would "be the powerhouse gateway to hundred-million-dollar business ventures in the Pacific." The idea was fostered by some of the region's most powerful political, business, and civic leaders.

In the Bay Area, it is not difficult to locate these would-be architects of the postwar city. The California State Reconstruction and Reemployment Commission, created to smooth the transition from war to peace, helped form what would come to be called the Bay Area Council. Nearly every large corporation in the area—Kaiser, Bank of America, Hewlett Packard, Lockheed Missiles and Space, Bechtel—had a representative on the council. Its machinations would have repercussions across the bay. The council would create catalytic regional institutions that soldered together the many competing municipalities around the idea of

San Francisco as the imperial headquarters of the Pacific Circuit. BART was its first great project.

The campaign to pass the local funding measure for the rail system was heavily backed by the Bechtels, whose company would go on to receive tens of millions of dollars of contracts to build the system. Executives from Bank of America, Wells Fargo, and Crocker National Bank all had a hand in supporting the initiative and also in making money from it.

It would be unkind, however, to suggest that these men and corporations backed BART *solely* because of the possibility to funnel money into their pockets. Economies worked differently back then. These were regional corporations whose executives could have reasonably felt tied to their hometown. For them, BART could lend that futurey sheen to their San Franciscan headquarters. Wells Fargo's advertisement in a special supplement in *Fortune* touting the project had a banner headline: "In 100 Years We've Moved 10 Feet." They were a San Francisco company and proud never to have left Montgomery Street.

In 1962, a bond measure to support construction of the system was put before voters in the area. After some complicated wrangling over which counties would join the system, the bond passed in Alameda, Contra Costa, and San Francisco counties. Oakland and SF were in, along with Berkeley and a host of smaller East Bay towns. It was not the bay-circling system that had been originally imagined, but it was still a huge deal. The $792 million bond issue and the billion-dollar price tag on the system made it clear that this was "the largest single public works project ever undertaken in the U.S. by the local citizenry."

Bill Stokes, the same Stokes who promoted urban renewal in Oakland, wrote a nine-part series called "The Future Is Now." Stokes argued the Bay needed to be seen as a "cohesive metropolitan area" that was rationally planned together. That was how cities, especially Oakland, could conquer "de-centralization—the 'flight to the suburbs,' as it has been called." His prescription, borrowed from the region's elites, was a regional transportation system "so cheap, so fast, and so efficient that it would be ridiculous not to use the facilities."

If you want to build a train system, you need to buy the land first. The task was gargantuan and nearly unprecedented. BART needed to secure 3,600 parcels of land from homeowners, businesses, and government entities right in the heart of the metropolitan area. And fast. During our era, where a fight over a burgeoning homeless encampment might last years, it's hard to imagine the government simply taking so many properties from people who owned them. But they did.

An obvious precedent existed for the project's real estate needs: the freeway network, which had been extending its tentacles in California since the mid-1940s, thanks to gasoline taxes and federal highway funding. As hundreds of millions of dollars flowed into the state from the federal government each year, California built thousands of miles of highways all around the state.

The "freeway" was the "Engineer's Answer" to the problem of traffic. It was also the military establishment's answer to the problem of defending a huge country with troops. And finally, freeways were the urban planners' answer to segregating people and activities that they believed should be kept apart. Freeways carved up and encircled West Oakland, the most important being the Cypress Structure, which cut Prescott off from the rest of the city, pinning it against the Army base and port. Hunters Point, the San Francisco neighborhood where the Black population was expanding the fastest, was hemmed in by Highway 101 and what came to be known as 280. The exhaust of thousands of commuters fell onto the residents, who were increasingly isolated from the rest of the city, just as planned.

Building these freeways took a remarkable financial and legal effort, staffed by people who became known as "Right-of-Way Men." They were the guys who bought up the deeds along a route. Juiced by billions of dollars of spending on infrastructure construction, condemnation law sprang to prominence. The law tilted heavily in favor of the state. California put out a pamphlet laying out the government's position, titled "13 Million People Want My Property."

Shortly after moving in, Amos and Clemmie Hunter found out what would happen if the state wanted a piece of their land. The freeway engineers realized that they needed to connect the Bayshore Highway, 101, with 280. That route ran right through their backyard, which was near the old railroad right of way. So the state sliced off a piece of their land, but the Hunters kept their house.

With the BART bond having squeaked through the vote, the new organization quickly became a real estate acquisition machine under John Daniels, a chain-smoking organization man. He ate prepacked sandwiches every day for lunch and his primary hobby was collecting rocks. Daniels lived just a mile from the Hunters in a two-bedroom home on the other side of the new freeway. The house was squat, a picture window framing a strip of grass between a driveway and the curving walkway to the front door. The man had spent forty years in the right-of-way business working his way up to earn that home. It epitomized good middle-class living, perched on the flank of a golf course in Daly City, which was 97.4 percent white in 1960.

Daniels had come from the California Department of Highways as the top man in Northern California. The state had purchased $300 million of property under his command—and he copied his previous organization's documents and forms nearly word for word.

He built his team from the Highways ranks, too, pulling people in from all over the state. "We liked to remind ourselves that 'under all is the land,'" one of his men recalled. Well, the land and the power of eminent domain as instituted through BART Resolution 305.

This operating rule delegated massive authority to the real estate division, allowing it to function without the interference or help of the board. This had several intended consequences. First, the real estate men were the only ones who dealt with the property owners. The board never voted to approve or reject individual purchases. This meant, as the right-of-way man Wally Mersereau recalled, that the board did not have to "deal with individual property owners making appeals of their positions." Second, Resolution 305 also made the real estate program less transparent. BART began hundreds of condemnation proceedings against different property owners who did not want to sell their land. This allowed them to take close-to-immediate possession of the properties, and toss the condemnation lawsuit resolutions into "executive," which is to say, closed, sessions of the board.

No BART board member voted against a single condemnation resolution throughout the entire original acquisition program. When BART asked for your land, it wasn't really asking.

This allowed the real estate team to deliver tremendous productivity. They had far less time than highway right-of-way men, but they never once slowed up the construction teams. They were proud of their work. They thought of themselves as fair men, who sincerely wanted to give citizens "fair market value" for their homes. Daniels instituted a novel policy of ordering up two appraisals for each parcel and offering the owners the higher number. What could be more fair than that?

Long before BART's men got to people's homes, residents knew what was coming. They could read a map. If your home was near the path, you knew there was a good chance you would lose your home. What you didn't know was when or how to approach this powerful new entity to get the best price. This complex battle was Margaret's first experience of community organizing.

She was dispatched to go door to door and tell the Wades, the Henrys, and all the other Ingleside families that there was a meeting at her mother's house and they had best attend. Their big home was the natural gathering place.

The neighbors would share information about what to do. Should they take the deals that they'd be offered? Should they protest? How? They'd been traveling for such a long way, sold across an ocean, marched west, held in apartheid, before finally escaping west during the war right back into new, clinical forms of segregation. And now they faced a new bureaucracy, which held itself innocent of all crimes.

Amid the turmoil, life went on, and even multiplied. Margaret ran track, and she liked to run the streets with her sister Paulette, too. They'd end up in Hunters Point. Hunters Point was not Ingleside, and girls like her could not usually hang with the kids who grew up in the housing projects, amid the deprivations of life there. But Margaret could. She was extroverted, feisty, and she loved going to parties with her brothers. She was also a bit wild. You never knew what she might end up doing. And one thing she did was fall for one of a band of brothers in Hunters Point, a musician and artist, Jeffrey Mayo.

Before she knew it, she was pregnant. The family held a baby shower. Margaret posed for a picture in a flowered dress and a copper-colored cape tied in a big bow around her neck, a long necklace draped over her stomach. Her hair is rolled into a little crown and big round glasses are perched on her nose. Her brother, Billie, drapes one arm around her, the other holding cigars, a black bowler hat tilted on his head. They were kids. In her senior yearbook, she listed one of her likes as "Homemaking." And then, on August 2, 1964, there was Antony L. Mayo in her arms. Margaret was a mother at seventeen.

In Ingleside, the first neighbors began to sell in February 1965. The Chatmons, the Rhodes, the Camachos. The Thompsons sold in March. The Eatmons in April. The Hugheses in May. The Hunters agreed to sell their home on June 10, 1965. It was document number O-M287, signed by Amos and Clemmie in their own hands, as well as by John Daniels in his practiced, looping script.

The appraisers didn't want to give them any value for the unpermitted additions that had been made to the house. They got $22,650, not too far from the median home price for California, though not as much as their parents had hoped.

Others held out, such as they could, forcing BART to condemn their homes. On June 24, 1965, the BART directors—each a white man—sat around a big conference room table at 814 Mission Street in San Francisco for their regularly scheduled board meeting. Among other business, they issued Resolution 551. In the meeting minutes, it received one, pro forma line: "Director Anderson, continuing, reported regarding one condemnation resolution required for the Outer Mission Line." BART began condemnation proceedings on at least twelve local properties. And just like that, their block received its final blow. BART had all it needed to begin construction on the Daly City station and the track leading up to it. Through that month, BART acquired 1,103 parcels of land.

Whatever Black people had, white people could take. That was the lesson of the first organizing Ms. Margaret ever did.

←---   ---→

In Oakland, a few blocks from where Ms. Margaret would eventually settle, the fight against BART was led by Beatrice Sneed, president of the West Oakland Homeowners and Tenants Association. As in Ingleside, the neighbors who found their homes targeted by the planning documents began to gather to discuss what they should do. While the right-of-way men saw a clean and fair process, the residents of West Oakland saw something else. Sneed contacted Governor Pat Brown, her U.S. representative, Jeffrey Cohelan, and State Assemblyman Byron Rumford. Most amazingly, nine days after Robert C. Weaver became the first ever Black cabinet member, as the first secretary of the newly created Department of Housing and Urban Development, Weaver answered a letter from Sneed, indicating his sympathy for the plight of Prescott.

"The President has asked me to thank you for your letter of January 13 concerning the dislocation of families that will take place as a result of the construction of the Bay Area Rapid Transit system," Weaver wrote. He went on to say that the federal government did not have jurisdiction, in a narrow sense, over BART, but that Oakland itself needed a "workable program" for housing, and that the city "has been informed that a relocation service must be provided for families displaced by the Transit system."

With this letter in hand and her political organizing gaining the attention of a newly formed activist group called JOBART, Sneed went to the Oakland City Council as Weaver had recommended to prevail upon them for help. She impressed.

"[Sneed] held the council near-spellbound last night with a quietly spoken complaint about how BART is obtaining residential property in that area," the deeply conservative Oakland Tribune wrote. "She charged that residents are being forced out of their homes by BART, receiving prices insufficient for them to buy comparative residences elsewhere."

A local publication that covered the poor areas of the East Bay called *Flatlands* printed the speech in full. Sneed said she resented the BART negotiators, who told residents they couldn't expect much money for their property because the neighborhood was "run down." But why was it run down, Sneed asked? Absentee landlords and a city government that had abandoned them.

"As taxpayers, in the past we have not cost our city a great deal of money. For we have not had good lights for our streets or good paved sidewalks to walk on without stumbling and falling because of their poor structure," she said. "We have not had good sanitation. The Police service has been poor, and many other factors I will not mention." In effect, the city's negligence had generated a discount for this other government entity.

Sneed then called into question the methodology of the appraisals. The BART appraisal handbook called for the valuation of a home based on the "highest and best use" of the property. So a piece of land that *could* be developed for something very valuable had to be appraised *as if* that were going to happen. In West Oakland, many of the homes had found themselves zoned for heavy industry, and as such, they could have been used for "higher and better" uses. So, Sneed asked, why were they receiving *residential-level* payouts instead of industrial ones?

Her other major point was that local residents could not "purchase comparable houses in our own West Oakland area with the prices BART offers." Whatever BART thought their homes were worth, the market for Black home buyers in Oakland was grim, in large part because of how much bulldozing the government was doing. "The citizens affected by BART bear grief. Anxiety exists," Sneed said. Where would they live?

Many were tenants, who were thrown out without getting a dime. *Flatlands* interviewed some of them. The stories range from terrible to tragic: One old man had been living in a room in West Oakland for eighteen years. When he lost his place, there was nowhere for him to go. A landlord sold a house out from under a mother of six. "The landlord stopped collecting rent around about November," she told *Flatlands*. "Guess she sold the place then. She ain't never tell us anything."

BART took possession but never made it clear that the family would have to pay the transit district rent. Then BART showed up to collect the back rent and told the woman she had to pay it all or move. "I guess we gotta move," she said, but was stalling for time because she didn't want to pull her little ones out of school and disrupt their schedules.

As urban redevelopment progressed, some scholars, technocrats, and legislators came to agree with Sneed that the systems that had pulverized America's inner cities were deeply unfair. A prominent economist named Anthony Downs wrote a paper called "Uncompensated Nonconstruction Costs Which Urban Highways and Urban Renewal Impose Upon Residential Households." Unlike local officials, Downs was able to accurately describe the moral economy that cities were using to revitalize themselves. "Fair" appraisals masked a system that was deeply unjust to poor Black inner-city populations. These residents had to make "essentially personal sacrifices" to help "the good of the public in general," Downs wrote. "It is therefore the duty of the public authorities concerned to compensate them for these sacrifices."

People in poor areas, where it was cheapest for governments to build new infrastructure, were *subsidizing* the rest of the city. In West Oakland, the breadth of the imposition was stunning: In an area that housed roughly 40,000 people in the early 1960s, the government destroyed between 6,600 and 9,700 housing units. Thousands upon thousands of people were pushed out of their homes; the neighborhood's population fell by 14,000.

Downs lays out twenty-two different costs that fell on people living in places like West Oakland during the urban renewal era. Some are universal and individual: disruptions resulting from the forced move, moving costs, the costs of finding another place to live. But others, the *whole community* had to bear: messed-up established routes and communications, reductions in the number and quality of businesses serving the neighborhood, general neighborhood disrepair, and the environmental degradation resulting from putting in polluting, noisy projects such as freeways and an overhead train.

And all this was happening primarily because of elite planning processes, drawing on the racist rootstock of American real estate. If rural southern segregation had its roots in the feudalism of old Europe, the new racism of San Francisco and Oakland had a modern tinge. It acted less on the individual Black body and more on the Black body politic. A new kind of "administrative evil" emerged. The social scientists Guy Adams and Danny Balfour coined the phrase to describe systems that submerge dehumanizing, terrible policies inside an envelope of technical rationality. "Ordinary people may simply act appropriately in their organizational role—in essence, just doing what those around them would agree they should be doing—and at the same time, participate in what a critical and reasonable observer, usually well after the fact, would call evil," they wrote.

Robert C. Weaver's *The Negro Ghetto* made the case in 1948 that government and private industry were working hand in hand to create segregated ghettoes in northern and western cities. Their combined efforts had destroyed Black economic fortunes and "condition[ed] Americans to accept racism and the rantings of demagogues who scream about the superior race, while supplying ammunition to the Communists in their never-ending campaign to win over minorities and disadvantaged people." Even within the Federal Housing Administration, there were racial relations advisers who could point out the injustice of the system. "Who are we kidding when we say, on the one hand, that minority groups 'prefer to live together,' and then proceed to utilize every device available in the market place to dictate that they do so?" Robert Pitts, the San Francisco region's racial relations adviser, asked at a conference in 1955.

As civil rights critiques and legal victories battered the walls of unreconstructed white supremacy, racism retreated further into numbers. "Quantification is a way of making decisions without seeming to decide," wrote the historian Theodore Porter in *Trust in Numbers: The Pursuit of Objectivity in Science and Public Life.*

Far from the conditions on the ground, mid-century planners created deeply unfair conditions in American cities, ones that would lead to the long urban crisis of later years. One was down, one was up—by design. But nobody within powerful institutions would own up to it. The Cold War–era U.S. government could not be seen as tinkering too much with the market. The interventions were framed as helping markets work better, rather than racially restructuring them. "The state was subsidizing white suburban home ownership and abetting racial segregation, all while insisting it was doing no such thing," David Freund wrote in *Colored Property*, another touchstone work.

And so, administrative evil became the moral essence of the mid-century process of urban redevelopment. The southern cry of "states' rights" became the western worship of white people's property rights, free from government meddling, tucked safely inside a fortress of numbers.

There was no winning, but that didn't mean you couldn't fight. Beatrice Sneed held out longer than almost all her neighbors. Most of them signed away their properties in 1965, '66, '67. She held out until February 27, 1968, a day of little note, just some space on the civil rights timeline between Huey Newton's gun battle with Oakland police and the assassination of Martin Luther King, Jr., in Memphis. Her former home lay underneath what is now the platform where you stand to get to downtown Oakland.

By the end of the decade, West Oakland's bustling business district was in shambles. In the memories of local residents such as the activist Oscar Carl Wright, the government's takings hadn't just destroyed homes, but economic cohesion. "Through eminent domain, they dismantled the black neighborhoods and black economics, black economics and ownership," Wright said.

People left in droves. You can read the history only in the empty streets, the decaying churches, fading signs (ESTHER's ORBIT ROOM), the grassy stubble of lots. Beatrice Sneed, too, fades into history. She reappears only in her 2001 obituary. She'd survived to eighty, nine years beyond her husband. She never had children. We read, "She loved music and traveling."

10

The Negro Family

After BART took their home, Margaret's parents searched for new housing, their problems deepening. Her father's work ethic and wandering eye kept him away from home. Her mother descended into alcoholism. The impeccably dressed young couple who had raised Margaret were aging into something sadder, tawdrier.

In January 1966, Margaret gave birth to her second child, Jarrod E. Mayo. His father, Jeffrey, was coming and going, sometimes out playing the congas with Santana, sometimes just out. When he was around, the young couple fought constantly and viciously, physically. He'd run the streets himself, but when she went out, he'd show up at parties and try to drag her back home. "Motherfucker, have you lost your mind?" she would say. "I don't even know why you thought you were gonna control her," a clucking cousin told him. Sometimes she thought he was on speed; other times just pissed. It was not good. There was too much trauma, too much drama, Margaret remembered.

Nonetheless, they got married that year, on August 30.

A month later, Hunters Point exploded in a full-blown urban uprising after police killed Matthew "Peanut" Johnson. Hundreds of people were injured. Nearly five hundred arrested. National Guard troops marched through the neighborhood.

It was the closest the Bay Area got to the sort of civil disturbance seen in Watts and Detroit. Many radicals believed that they were in a moment of revolution, the first steps in a war for the soul of America. Peter Levy, examining the period from 1963 to 1972, has identified what he calls the "Great Uprising" to describe 750 revolts across the

United States. Nearly every American city with a sizable Black population experienced something.

As related by the Hunters Point native and historian Aliyah Dunn-Salahuddin, the uprising was a "turning point" for the emerging radical movements of the Bay Area. Two months after it ended, Huey Newton and Bobby Seale founded the Black Panthers in Oakland. Two years later, San Francisco State students, led by the Black Student Union and Third World Liberation Front, struck the school, setting off a string of explosive protests across the country that lasted for years.

Margaret was not a key player in these struggles, but she knew the struggle. With the two kids, she moved to a new public housing project, Alice Griffith, known then and now as Double Rock. It was built in 1962 atop razed war housing, on landfill at the edge of Candlestick Park; the street plan was one huge cul-de-sac, like an industrial-strength parody of suburban life. Thirty-three low-slung two-story buildings faced the bay, 250 units of housing built as cheaply as possible.

Alice Griffith, though only four years old, was already short on hot water and trash collection. The local newspaper *The Spokesman* reported that "rats and roaches [were] taking over in Alice Griffith." They'd emerge in the evening from the sewers, scaring children and mothers alike. "It is a very shocking experience to step out of your door and see about ten rats running back and forth," the paper continued, "and then have one run into your apartment."

There was always something cropping up with Margaret's children. They got sick a lot. One day all three of them were recovering from trips to the hospital when Mayo showed up, grabbed his stuff, and disappeared. She'd been born into public housing, but she'd spent her adolescence not just with a roof over her head but a real home. And in the span of just a few years, her family had fallen apart, their home had been taken by the government, and she'd become a single mother living in the ghetto.

White American society had many thoughts and opinions about women like Margaret. The Moynihan Report burst into public view in August 1965 with its finger-wagging stance toward "the Negro family."

Written by Lyndon B. Johnson's assistant secretary of labor Daniel Patrick Moynihan, *The Negro Family: The Case for National Action* argued
that civil rights legislation would not be enough to make people equal.
But, as Moynihan would have it, the problem wasn't just racist white
people but the culture that Black people adopted in response to its bru-
talities. Black families were beset by a "tangle of pathology" that was
"the single most important social fact of the United States today." The
decline of the institution of marriage among Black families had led to a
second problem: a large proportion of households that were headed by
women. The charitable interpretation of this concern was that a labor
market structured to provide men the best jobs meant that woman-
headed households were doomed to economic problems. Moynihan's
answer, though, was not to expand opportunity to women of all racial
backgrounds but to tailor the design of jobs programs to get Black men
employment.

The Moynihan Report was one of the more subtle arguments about
Black women's role in their families and communities. A much uglier
version emerges in *Hunter's Point: A Black Ghetto*, by Arthur Hippler, a
white anthropologist who did his fieldwork during the time Margaret
spent in the neighborhood. Hippler concentrated on the projects up on
the ridge, just above the shipyard, across the basin from Double Rock.

To Hippler, Black women were a—and maybe even *the*—problem
in the ghetto. Hipper says that he used to think that Black people's "individual personality problems" were "a function of the racial deroga-
tion by whites, and that solutions to this problem could be found by
viewing it as essentially structural." But, after working with Black and
Native American people, he decided that "there are fundamental differences among human groups in their ability to achieve and create." His
work, he noted, is "essentially negative," and offers no "ameliorative"
suggestions.

"I made no pretense of somehow truly 'understanding' the quality
and meaning of life in Hunter's Point as it is felt and experienced by its
inhabitants," he wrote. And yet he did feel comfortable contradicting
the judgments of the objects of his study. One woman, whom he names

Mrs. Sincero, confesses that she's trying to get a divorce. "My old man, he beats me too much," she told him. "He's scared of me, but I'm afraid he'll kill me."

Hippler, however, calls this her "personal pathology." "In reality," he goes on, "Mr. Sincero is a rather quiet, passive man who can be goaded into physical violence only on occasion. Mrs. Sincero does not seem to be aware that she apparently solicits the physical beatings she gets and yet uses them and her own manipulative attempts to thwart her husband sexually." For Hippler, Mr. Sincero's violence against his wife is *her fault* because she "has helped to induce, or at least to reelicit" his "sexual failure."

If Black women were a problem in families, they were an even bigger problem within the community. Hippler recognized that nearly every community organization was powered by "women who received their training in these informal associations of matriarchs." But rather than seeing this as a strength of the community, and these women as assets to Hunters Point, he sees this as perpetuating a pathological tendency to "female superiority" in which "male worth" is "deprecated." By the end of his book, Hippler has worked himself into painting the grassroots organizers in the community as "a real obstruction to any possible progress."

This was not how Margaret saw these women. They were, instead, powerful figures who drew her into community. Right there in Double Rock, Margaret learned at the knee of Bertha Freeman, who was, as a local Black newspaper would have it, a "fiery lady with a big heart." To be in Bayview Hunters Point during the late 1960s was to see many different versions of leadership in the community. A set of imposing women—Freeman, Julia Commer, Elouise Westbrook, Ruth Williams, Osceola Washington—came to be known as "the Big 5," and they both fought and worked with city, state, and federal officials for decades.

Their commitment to their neighbors was serious and long-lasting; they spent their entire lives trying to make the area a better place for everyone. They showed her how to fight the system from the inside and outside, dragging her to meetings all over the city. When Senator

Bobby Kennedy and other officials came to Hunters Point for a hearing after the uprising, it was Washington who told them, "If you are from Hunters Point, you get no chance at all."

These became Ms. Margaret's role models and some of her first mentors. "I'd go to all these meetings with the older ladies. I'm twenty-something and they were in their forties and fifties and sixties. Them ladies was tough. Them ladies had been fighting to upgrade Hunters Point when I got there for twenty, thirty, forty years," she recalled. "The whole thing was, 'You a youngster and you wanna come with us. Your thing is to take notes. And after the meeting, we can all go get our bottles and you can tell us [what you think]. Then you can ask the questions.' The whole thing was about giving me a foundation."

Over the years, the Big 5 grouping changed and it wasn't always five people, but the *idea* that there was this team looking out for the interests of the neighborhood mattered. Westbrook lived into the 2000s and was the sort of brash, funny leader that Ms. Margaret became. She sprinkled "baby" into her speeches, was determined and knowledgeable, and could hold her own with any part of the power structure. She's most famous for leading a delegation of Hunters Point folks out to Washington, D.C., to secure funding for rebuilding the dilapidated housing on "The Hill" in Hunters Point. Later in life, she ended up on many city committees, and when she died, power brokers from all over came to celebrate her, from Dianne Feinstein to Willie Brown.

Ruth Williams, for her part, ran a sex education program, largely for young women like Margaret, who hadn't really understood "rubbers." She was politically engaged, and saw her work with Planned Parenthood as part of the liberation struggle: she received letter after letter begging her to lay out the options to avoid having more children. Women wrote things like: "Is there any way to keep from having my seventh baby? I'm pregnant and I don't want another child. We have too many children for us to care for already and my nerves are pulled so tight that I'm afraid they will snap in two." And: "I've had a baby every year since I've been married. Right now, I just don't want to get pregnant again. I love my children and want to raise them right. How can I get birth control?"

While people like Hippler were presenting this idea of women popping out babies and castrating their men, these women were struggling
to stay empowered and alive. That was Margaret, too—just hanging
on. She made some bad decisions, as she acknowledges, but it's hard
not to wonder how much of her difficulties had to do with the negative
impact of the American government on her life.

By the time she was twenty years old, every place Margaret had
lived had been deeply impacted by the state.

The place she was born? Bulldozed.

The place her family purchased in Ingleside? Bulldozed.

Her home in the projects? Already falling apart and filled with rats.

PART II

The Revolution That Came, and the One That Didn't

Brian McWilliams

11

The Potentialities Are Beyond Imagining

Containerization was and is the linchpin of the Pacific Circuit. The loops and processes that send goods steaming from east to west did not exist in their present form before containers. The container revolution transformed longshore work, the environment of West Oakland, the economy of Oakland, and, over the decades, the entire Pacific Rim, as Asian countries raced to build ports that could send goods to Oakland (and later Los Angeles and Long Beach). Containerization set off a spiral in the scale of infrastructure that's centralized shipping between dominant ports. As businesses figured out how to use the whole new logistics system more efficiently, it sent transportation costs plummeting. Marc Levinson notes in his defining history, *The Box*, that the implication of this new technological network is clear: "Globalization, the diffusion of economic activity without regard for national boundaries, is the logical end point of this process."

At first, containerization was used for trade across the Atlantic, because the relationships between European countries and the United States were the most valuable on Earth. But that did not last long. The routes from the Asian countries to the West Coast began to play an ever larger role in the global economy. Ports on both sides of the Pacific cultivated these relationships, underwrote the costs of new facilities, and competed with one another. For the whole globe, these details might not have mattered, but for those on the shoreside of this new system, it could be thrilling and wrenching.

So much of the world that I've grown up in resulted from processes that began in the late 1960s and reached what might be their peak in the 2010s. There's no better place to trace that transformation than down Seventh Street by the old Army and Navy bases.

During the early part of the era, the Bay is famous for two entirely different reasons. Most obviously, it was one of the most important centers of the anti-war movement. The Third World Liberation Front, Berkeley peaceniks, San Francisco hippies, East Palo Alto Mexican radicals, and the Black Panthers were all laying their hands on different pieces of the military-industrial complex's connections and ideological underpinnings. They were connecting up the ways that people of color across national boundaries were being oppressed by what they saw as American imperialism. They resolved to fight it not just in Vietnam but at SF State and Berkeley, in Chinatown and in West Oakland.

At the same time, the first semiconductor companies were putting the silicon in Silicon Valley. They were some of the most dynamic companies in American history, poised right at the edge of science and technology, funded extravagantly by different arms of the government, and then exploding into commercialization. Intel, Hewlett Packard, and soon Apple, Atari, and all the rest were there.

The technology cluster that is now the envy of the world was taking shape south of San Francisco, but it moved quickly to take advantage of Asian labor. Fairchild Semiconductor, the company that launched dozens of spinoffs (including Intel), began contracting the assembly of some of its products to a Hong Kong factory in 1961. In the United States, it had to pay an assembler $2.80 an hour. In Hong Kong, the going rate was $0.25. And it wasn't just low-level employees but supervisors and engineers. Everybody was cheaper in Hong Kong. By 1966, it had hired four thousand people, and had expanded to South Korea. The whole flotilla of Silicon Valley companies that became known as the Fairchildren carried that business know-how with them.

It might be tempting to see the radical liberation movements, electronics takeoff, and containerization as unrelated, but they were not. The Vietnam War is what launched transpacific containerization. But it took off in Oakland precisely because the port was squeezed between the military and a poor Black community. Semiconductors went into radios *and* weapons. A computer could calculate efficient bombing runs or shipping routes. Labor exploitation could happen in San Jose or in Seoul, and as later activists would put it, those workers were on the same production line. The United States was binding its allies with technology and trade, while ignoring the collateral damage to local people and ecologies all over the world.

The analysis that developed among Bay Area radicals during that time created a potent mix of anti-imperialism, anti-racism, and anti-war sentiment, drawing on the power of the diversity of its participants and their experiences at the receiving end of American power at home and abroad. While they became culturally influential, the material infrastructure constructed by their opponents ushered in a new era of global relations. Corporate power augmented and even replaced direct U.S. authority. Machines became cybernetically wired into human systems of exchange. Everyone could use and understand the old Mexican lamentation, "So far from God, so close to the United States."

The system had powerful effects onshore, too. Capitalists who controlled production processes became fabulously wealthy. Regular working people's incomes stagnated, only partially offset by the falling prices for many consumer goods. Oakland and places like it entered a period of sustained decline.

For baby boomers like Margaret, the result was an adolescence filled with radical hope and new ideas, followed by decades of stagnation and struggle.

The Stanford Research Institute sponsored a series of symposia on automation in the 1950s, bringing researchers, industrial giants, and military heavyweights to San Francisco to hobnob. All automated processes, said one attendee, were simply manifestations of the new ways that "information, energy, and materials" could be rewired by "the elimination of human operators from great areas of production activity."

As factories had been around for decades, SRI conference attendees did feel the need to debate whether automation was really *new* new or simply a spin on Fordism. They were convinced, though, that this was not an incremental change on the horizon. "It may be true that there is nothing <u>basically</u> new about automation," a defense contractor told the group, "in the same sense that there is no <u>basic</u> difference between an atomic bomb and an old-fashioned TNT explosive."

Automation was what happened when you married new electronics and powerful machines to managerial optimization. It built new systems of production that transferred power from the people who worked to the people who built the machines that did the work. "Human operators" were rarely *eliminated* so much as disempowered and put under the control of new systems.

Containerization is a paradigmatic example of real-life automation. For centuries, the hulls of ships had been the "container." In break-bulk cargo, dockworkers stowed cargoes inside different parts of a ship. The nature of the work gave the longshoremen some power: they had to do a good job or the cargo could be damaged. If they were treated poorly (or even if they weren't treated poorly), they might also skim a bit off the top—pilferage, it was called. The box changed all that, massively reducing the need for waterfront labor and sapping the autonomy from the remaining workers.

One longshoreman-poet, Gene Dennis, captured the despair accompanying the changes to the work in a poem about automation.

> *My soul has been sucked dry and suffocated,*
> *By the shadow of a 40-foot container*
> *Restored by outrage at the mindless technology unleashed*
> *By cash register computers*
> *So Logical, So Methodical*
> *Casting aside bent bodies with poisoned lungs to proceed with greed*
> *So technologically correct*
> *A heritage caved in by the ponderous pounding of some*
> *Psychotonic, robotronic beast*

Technologists did not make mini robot longshoremen carrying things on their backs but rather they built an entirely different system requiring far less labor. That's how automation usually works.

Most containerization stories focus on Malcolm McLean and his company Sea-Land. The story of "a trucker"—really the owner of a very large trucking company—getting into the moribund shipping industry and revitalizing it with this one simple idea is irresistible.

But in the Pacific, container shipping began as an operations research project, conducted by a Johns Hopkins professor, Foster Weldon, with the aid of a new computer. He published the first paper explaining how to systematically containerize in the journal *Operations Research*.

Weldon was working for Matson, a major player in Bay Area shipping, thanks to the Caucasian aristocracy of families who ran sugarcane plantations (and much else) on the Hawaiian Islands. They, too, were known as "the Big 5." Matson was a sort of royal corporate child, then, the maritime infrastructure that the families who ran Hawaii decided they needed. Matson had a sibling, too, the sugar concern C&H (California & Hawaii), which built and operated a massive refinery on the Carquinez Strait, right where the bay begins to turn into the delta.

Matson was born as a monopoly and has remained that way, or close to it, for the last 140 years. Into the 2010s, the company maintained close to 70 percent market share on Hawaiian shipping, aided by the Jones Act, which requires that maritime trade must be conducted on U.S.-built and -flagged ships, which international cargo shipping companies do not own.

The monopoly profits created an opportunity for the company to undertake a serious research program into the future of shipping. This was the fashion of the mid-century, after all. The field was flush with success stories from World War II, and had taken immediately to computing.

At that mid-1950s moment, Weldon noted that "the containerization of general-merchandise cargoes is under active investigation by almost every major railroad and steamship operator in the country." The problem was how to make all the decisions that might go into designing a containerized cargo *system*. How could the profitability of container shipping be evaluated without . . . building out the whole thing? And what if every company built their own? Then they wouldn't be able to interoperate and many of the hypothetical efficiencies that could be imagined would evaporate. All the companies, Weldon said, had "their own pet theories on the detailed equipment requirements comprising a 'best' container system."

So Matson went to Weldon. The company was in an ideal position to build out a containerized cargo system, owning terminals on the West Coast and in Hawaii and the ships that ran between them. Weldon modeled the different possible cargo systems to help show which size container, for example, might be most cost-effective.

As important, his analysis also captured what many histories of containerization neglect: while it might seem obvious that shippers could capture vast cargo-handling cost savings by eliminating longshore labor, it was not clear the *full system* cost would be lower. Matson's Hawaiian trade was imbalanced: more containerizable cargo went to Hawaii than needed to go back to the mainland. And the cargo was not in the form of full container loads, so Matson would have to stuff boxes and then unstuff them on the other side. (This would be borne out in the real world, too, as it took quite a while for real-world shipping costs to decline in line with the labor savings on the waterfront.)

What containers most certainly *would* do, however, would be to weaken longshore labor's position. This would not be a mark against it in the Matson management world. After all, the ILWU had unionized many different parts of the Hawaiian economy, and its collective power was a serious check on the Big 5.

Matson began to invest in containerization at least in part because of the research program and Weldon's findings. By the time the paper was published in 1958, Matson had "a full-scale prototype" container system and had "announced a multimillion-dollar program" that would "result in the first large-scale schedule van-cargo service in trans-Pacific operations."

Matson also worked out key parts of the operation, such as the specialized shoreside crane, which is now the unofficial mascot of Oakland, and other container-handling equipment. In 1958, Matson commissioned Pacific Coast Engineering Company to build a 260-ton beast standing 113 feet high at a new terminal in Alameda. It could load a 40,000-pound box in three minutes over and over and over, without ever getting tired or straining a back muscle or needing to piss. The design PACECO created for Matson went on to dominate the container shipping industry. Hundreds were operating all over the world within a couple of decades.

While they were initially located in Alameda, they quickly outgrew their space on the waterfront, and signed up with the Port of Oakland. "Containers had grown so much faster than they or we had even dreamed of," said Ben Nutter, the legendary director of the Port of Oakland during the period. Oakland's ascent from sleepy backwater to globally significant port had begun.

Of course, Malcolm McLean's Sea-Land *is* an important part of this story, too. He was able to get a successful container operation up and running to and from Puerto Rico, especially after his competitors collapsed for various reasons, and he could feed on the kind of Jones Act–induced monopoly profits that powered Matson. As Sea-Land scaled up and the business stabilized, the company's rates dropped, effectively lowering shipping rates for the whole island.

McLean and the engineer Keith Tantlinger created several important innovations, such as the twist-lock system that allowed containers to be stacked like Lego bricks. The questions longshoremen answered every day about how to stow cargo in holds could be answered not just for an individual ship or a single hold but for all ships and all holds and all boxes.

The ILWU could see the writing on the wall. Longshoremen had been one of most radical labor groups in the west since they'd struck in 1934. There were a lot of them, as unloading ships required a lot of labor, and they stuck together. They'd proved their might with strikes and negotiations.

It wasn't just *a job* but a way of life. Dirty, dangerous, backbreaking, disrespected, but also challenging, flexible, social, and *fun*. The docks were filled with storytellers who drew on the common heritage of the profession to craft jokes and commentaries from their very specific position in the city's hierarchy.

One San Francisco conduit for the waterfront's stories was a hilarious, cantankerous drunk and poet-novelist named George Benet. He described himself, by way of a friend, the Beat poet Lou Welch, as "a short, fat, alcoholic longshoreman from the Mission District. A cigar-smoking, horse-playing son of a bitch. And worst of all, an incurable romantic." To hear people tell it, he was all those things and better and worse. One of his favorite stories went like this:

> We wore a uniform, a light cap, hickory shirt, and frisco jeans . . . At one time, probably 95 percent of the men on the pier had this uniform on. And there was a man working on the docks. His name was Strong-Arm Louie. He was a very powerful man when he was younger. He became more subdued when he was older and he's dead now, god rest his soul. Strong-Arm Louie was a toilet paper stealer. Every day that he left the docks, he'd take two or three rolls. I always envisioned him in the Mission District. He never married. He was living with his Irish Mother, and while they're having supper, down in the basement, there are about 25,000 rolls of toilet paper. What the hell would he do with all this toilet paper?
>
> Strong-Arm Louie is riding the Mission District bus and he's sitting near the back and the bus is crowded. It's full. This is before woman's lib and a woman gets on the bus. As she goes to the back of the bus, she hesitates before all the men and nobody gives her a seat. But when she gets near Strong-Arm Louie, he gets up and gives her his seat. He's got a big cargo hook and the cap and the frisco jeans and everything on. She can't let this slide by.
>
> In a very loud voice, before she sits down, a very loud voice, she says to the rest of the bus, "Well," she said, "at least the long-shoremen are gentlemen." And Strong-Arm Louie tipped his hat and put it over his heart, and he said, "You're fucking right we are, lady."

In living memory, the longshoremen had had pitched street battles with police and their associated toughs, seen their leader persecuted by the federal government, watched their friends die, been routinely injured, and read their position misrepresented and twisted by the big city papers. Nutter, the Port of Oakland executive director, recalled that if the port interests had political trouble, they'd just visit Bill Knowland, publisher at the *Oakland Tribune*, "and talk about a problem, and we'd get support in the newspaper."

The longshoremen knew what it was like to be on the receiving end of power. Their counterparts in New York were controlled by mobsters, dragged into working for the Mafia, and generally caught in the underworld, whether they wanted to be or not. Many longshoremen pilfered many things from the cargoes they handled. If you could ship it, they could find a way to steal it. But the way many of them saw it, what wasn't stolen, one way or another, in the wider world?

Benet was a fixture at the bars that dotted the waterfront, clustered around the pierheads where longshoremen and other people who worked on the docks gathered before, after, and during the workday. His favorite was the Eagle Cafe, which stood at Powell and Embarcadero, but anywhere there were drinks was also his favorite.

"So many people loved George and at the same time couldn't stand to be around him," recalled Michael Vawter, a member of the younger generation who joined up in the '60s. The longshoremen's wives barred him from their houses. He'd go on horrible benders, caught in the memories of a life he'd spent ending up on the waterfront. His writing is about not misery but mystery and memory, the surprise of having survived so long.

In one passage from his semi-autobiographical novel, *A Short Dance in the Sun*, the main character's gang is working a banana boat. There's Borrego Red, Sweetdick, Siberia Sam, Rankle John, Roger Walsh, and a Dutchman named Roebling, a multicultural, multiracial bunch possible only in ILWU Local 10. Down in the hold, taking the huge banana stocks and loading them onto the belt that hoisted them upward, Roebling hurts his arm. The men find him dead outside a cafe, meaning he wouldn't get a $25,000 death bonus for having died aboard. They cause a distraction and sneak him back on, tossing him down a ladder, then swear he died from the fall. Roebling had five kids, after all.

Then the men in the novel sit across from the Palace of Fine Arts, drinking gin and tonic from bags until they can't feel anything, as the city's evening chill descends on them. "Wherever you were, particularly at the racetrack, [George Benet] would seek out the one person who was the most depressed, lonely, broken, broke, having lost everything, just bottom of the barrel," Vawter remembered, "he would always go up to them and include them."

The point is: there was this whole culture, even a natural politics, that grew out of the work. A longshoreman spent his time with other people. Partner, gang, local, union. These were complex social relationships, but they were one of the main appeals of the job. You got to know guys. They understood you. Not everyone had to be best friends, or even to like each other, but there was a bond. Some of it was that dockwork tended to run in families. Their fathers and uncles had been longshoremen, grand-fathers, brothers, cousins, people from the old country, brethren from down south. But it wasn't an insular community. Far from it. This was a distinct kind of cosmopolitan worker who was used to interacting with people from all over the world. The longshoremen themselves contained multitudes and they had purpose: keep the hook moving, protect the union, don't get killed, master the art.

Work came and went, but the life of the longshoreman remained. You went to work only when you wanted to: if you wanted to go fish-ing or to play the horses or were just too hungover to go in, you just didn't. That freedom meant something, maybe everything, to many of the men.

The ILWU leader Harry Bridges could see that containerization would be more like an atomic bomb than TNT. Automation would bring not just a loss of jobs, eventually, but the restructuring of the wa-terfront away from the humans. All the stories and the drinking and the pilfering, the irrationality of longshoring, the way it felt old and weird and uncommitted to managerial efficiency: that would get ironed out.

It would also do away with the truly brutal work that the men en-dured, which wreaked havoc on their bodies and minds, leaving them to lean on alcohol to kill the different kinds of pain. For, as much as many men loved the waterfront, there is no denying the work was lit-erally backbreaking.

So Bridges took a deal. In 1960, he worked out the Mechanization and Modernization agreement with the Pacific Maritime Association employers. It came as quite a shock to the rank and file. The M&M was immediately (and has been for all the years since) a polarizing agreement. Older men would get the possibility of retiring early. The union would "shrink from the top," keeping younger men around, but tightening up the inflow into their ranks. They'd be smaller, but richer. The deal flipped Bridges's reputation with the very people who'd despised him for years as a Communist. Suddenly, he was a farsighted visionary.

On the one hand, labor got, as they put it, "a share of the machine." For every ton of cargo that got moved, some money went into a fund for longshoremen. The ILWU had used its negotiating muscle to extract *something* from the employers at the very end of the tenure of the PMA negotiator whom the union respected and trusted, J. Paul St. Sure. The longshore workers would stay central to whatever new system was constructed. And for those who got squeezed out, there would be a dollar-padded landing to do other things. It was a hell of a lot better deal than most unions got for industries going through similar things, then or now.

On the other hand, the M&M introduced mortality to traditional longshoring. It would be years before the actual work went away and everything became containerized—a generation of longshoremen would live through the transition—but the M&M was a recognition that eventually, the old longshore life would be over. Men would work in and around huge machines. The pace and rhythm of the job would be determined not even by some*one* not in the union but some*thing*, the machine. The gangs would be broken up. Everything would move faster. Most of the work would require significant skill with specific machines, which meant that more men would be hired directly by companies as "steadies," rather than through the hiring hall. The job would become more like others.

In San Francisco, the bawdy, rowdy street cosmopolitanism of the longshore would retreat. Warehouses would become lofts, piers would become offices and museums.

In Oakland, containerization would mean explosive growth for the port. A sleepy spot would be among the most important container ports in the world in the ten years after Harry Bridges signed the M&M.

In West Oakland, containerization would mean trucks by the thousand idling in and rumbling through the neighborhood, pumping diesel exhaust into the air.

←----   ----→

Imagine yourself on a freighter finishing a long journey across the whole of the Pacific. You left the Oakland Army Base weeks before, passing under the Golden Gate Bridge, skirting the Farallones, and then charting a great arc south and west west west. Another great bay awaits through another marine gate, flanked by rocky escarpments. In the distance, more jewel-toned peaks.

On a peninsula reaching down from the north, the American military has leveled and paved itself some territory. Piers stretch out into the water. An airstrip waits for more new recruits and to send the injured home. Sand blows around feet. For some soldiers, this will be the first or last piece of Vietnam they see.

At most of the piers, longshoremen unload old naval cargo ships bristling with cranes. Gangs work their way into the holds, signaling crane operators to hoist up the contents to the docks, pallet by pallet, the old way.

On one pier, though, two shoreside gantry cranes perch, tiny by current standards, but still huge. Their steel ligature is dark gray, but the engine house that powers the lifting of boxes is bright white, like the tops of the puffy clouds that dot the sky. The base is marked with Sea-Land's black-and-red logo. Standing there, on Pier 4, on this spit of land more than six thousand miles from the west coast of America, you could look out across the beautiful water, all the gradients of blues, over to the hills, carpeted in green, the first glimpse of the jungles that have absorbed your friends.

Trucks wait to carry containers to what passes for a yard. They, too, are Sea-Land's. The military business allowed the company to build out a whole transpacific system for containerization on the government's dime, improving operations as it grew and catalyzing the growth of the new form of shipping. This beautiful place, Cam Ranh Bay, was the beachhead not just for vast amounts of Army cargo but for containerization itself.

On the other end of the circuit, Oakland had remained the center of the military logistics system for the Pacific Theater. As it had been during the Korean War, it served as a way station for new recruits being sent to Vietnam, who slept in warehouses filled with cots before boarding jets to Southeast Asia.

On land abutting the base, the Port of Oakland had leased a terminal to Sea-Land in 1962, what was and is called the Outer Harbor. Sea-Land began running service back and forth to the East Coast. With an up-close view of Sea-Land's and Matson's early successes, the port's executive director Ben Nutter decided to bet heavily on containerized facilities. San Francisco's port director, Ray Watts, was much more skeptical, telling Nutter that containers wouldn't last. "We've been handling cargo like this for fifty years, and we're not going to stop now," Nutter remembered him saying.

The thing with containers, though, is that they require land, lots of land: dozens of acres for each terminal and then warehouses, too, preferably nearby (though that changed over time). With San Francisco real estate prices, more and more companies began getting warehouse space in Oakland. The freeway access was better, the traffic less intense. It was easier to get to the interior of the country, and there were rail lines there, too.

The Port of Oakland wanted more land. In fact, it wanted to build an entirely new terminal, out at the end of Seventh Street. The problem was that the Bay Area's politics had turned against filling the bay, a crucial early victory for traditional environmentalism. The Save the Bay movement had been launched by three powerful Berkeley women, led by Catherine Kerr, the wife of the president of the University of California. They were horrified by Commerce Department projections that the bay could be mostly filled in by 2020 and began to crusade against these kinds of projects.

The bay had to be saved because, as one chronicler put it, the waters were "of such unforgettable beauty as to be of national significance." Within five years, the environmentalists had gotten a new state agency created, the San Francisco Bay Conservation and Development Commission, or BCDC, as everyone calls it.

The formation of the new agency and the creation of the Seventh Street Terminal were on a collision course. If the port didn't start construction before BCDC's jurisdiction along the shoreline came into effect, the project would probably be tied up for years, if it happened at all. "We knew that if we waited beyond that date, we'd have to go through all the trouble of hearings and all that sort of thing with BCDC," Nutter said. "So we did everything we could to get it approved in Washington before that and got the work started, so that we predated their authority."

The on-time start allowed Nutter to cut a deal with BART, whose trains emerged from underwater on the north side of the terminal. In exchange for that easement, it contributed the dirt generated by tunneling through the Oakland Hills to build the massive new expanse of land. The *Tribune* devoted an entire article to this process, headlined, "It Took Lots of Fill." By the port's reckoning, 13.5 billion pounds of dirt went into forming that extension of the port, which is now named after Nutter.

As the port was being built, Sea-Land became increasingly aggressive in the Pacific. "In the next 20 years, all significant trade routes will either containerize or be killed," the company's Harry Gilbertson told the *Oakland Tribune* in 1966. He was right.

The local newspapermen breathlessly reported the growth of their port, which seemed to augur the day that the East Bay would finally become the dominant force in the Bay Area. "San Francisco docks are so congested, so clobbered up, that no container operation with half a mind would go near it," Gilbertson said. "They'll come to Oakland."

Sea-Land, from next to the Army docks in Oakland, pitched the military on building out a container system across the Pacific. As Marc Levinson, author of *The Box*, tells it, the Vietnam War effort was having a hell of a time logistically. "There were a few docks along the Saigon river and that was pretty much it. So this raised the question of how do you feed and equip an army in a place that has no facilities to transport cargo or to import cargo," Levinson said.

Instead of loading cargo into their own supply ships, the Army began to contract with Sea-Land. Oakland became "the largest military port complex in the world."

That was only the beginning, though. The U.S. government paid the full cost of a round-trip voyage of Sea-Land ships, but they sailed home from Asia with mostly empty containers. Malcolm McLean cut deals to begin stopping in Japan, as his competitor Matson already did. Japanese shipping lines caught on quickly, spurred by visits with Nutter and other American port officials. By the time the Seventh Street Terminal opened, it immediately became home not just to Matson but to a flock of Japanese shipping companies, carrying the goods of the 1960s' fastest-growing economy. Asian electronics companies were quick to containerize, and an already strong connection between the Bay Area and Asia strengthened.

Sea-Land kept feeding on military contracts. By the peak of the Vietnam trade in the early 1970s, the military work probably accounted for close to half of Sea-Land's business, Levinson told me.

The Port of Oakland was flying high, too. In 1957, the port had handled 2.5 million tons of cargo. By 1972, containerization had tripled those numbers, with Nutter receiving much of the credit for his vision and insensitivity to community concerns. His successor as port director, Walter Abernathy, saw Nutter front-running the new regulations as a downright heroic act. "He planned the Seventh Street Marine Terminal on 140 acres of new fill at the time public and political opinion had coalesced firmly in opposition to such projects. He started construction hours before the deadline when new state regulations prohibiting such construction became effective," Abernathy remembered. "The public, news media, and most elected officials demanded that the project be stopped and subjected to the new bay fill regulations. He and the port were branded as insensitive environmental brigands."

They were, in fact, insensitive environmental brigands, no? But it's also important to recognize that they were acting as they had been acculturated into public service. Listening to the community was *explicitly not* in their remit.

It was, Abernathy admitted, "a different era," and one that he clearly longed for, "when the public entrusted more of its major public works to the good intentions and judgments of professionals. No environmental impact reports; few permits, little regulation; and limited litigation."

If you were to trace the relationships that the city of Oakland had in those years with all the points of the Earth, there would have been many lines moving around the Bay Area, Northern California, down south. Flows of people and products moving up and down the West Coast, out to Hawaii, into the Central Valley. But the most important links for the development of the city began to cross the Pacific. After the Japanese set up facilities, Korean companies followed suit. Then Taiwanese and Chinese companies. The long-awaited connection between Oakland and the Pacific was established and energized.

Containerization would have happened without Vietnam or Oakland, but as with many Cold War innovations (rockets, transistors, computers, the internet), container technology was pulled forward in time by the U.S. government's needs and willingness to spend whatever it took.

It's telling that nearly no port histories or officials of that era mention the impact of increased container traffic on the adjacent neighborhood of West Oakland. And it was the relative powerlessness of that community that made the port's explosive growth possible. What sat behind the piers in San Francisco was extremely valuable downtown real estate. There was no land available to store the containers that were waiting to be loaded on or off a ship. How could vast numbers of container trucks get in and out of the Embarcadero? Where would they idle?

But what was near the Port of Oakland? Only West Oakland, a poor Black neighborhood that had been slated for destruction for more than thirty years as a matter of official policy. If the diesel trucks serving the port happened to pour particulate matter into the lungs and brains of the neighborhood's children, that was just how the world worked. West Oakland simply had to bear the brunt of progress, whether it was BART running down the main strip, urban renewal tearing out acres of housing, or a shiny new port next door.

12

Black Power and the Third World Movement

In the late 1960s, Margaret was still living with her children in Double Rock, helping out with Head Start, the new pre-K program that gave kids from poor areas a chance to acquire the skills they would need to succeed in school. She and her kids were always getting sick. Tony and Jarrod had the childhood rashes and everything else, but also asthma, like so many kids raised in places with poor air quality.

Head Start was one of many programs enacted under Lyndon Johnson that became grouped together as the "War on Poverty." In the decades since, this policy push has been used as a punch line and cautionary tale about the failure of government, but in ways both intended and unintended, it had a huge impact on the Bay Area.

For its entirety, the War on Poverty's programs came under attack, especially the more controversial parts like the Community Action Programs, with their emphasis on "the maximum feasible participation" of the poor in planning and decision-making. Black activists and political leaders took that phrase and used it as a lever to crack open local power structures. This was one of the *intended* effects of the War on Poverty. Federal leaders thought that urban political machines and charity organizations tended *not* to help the people they were supposed to serve.

In many cases, like Oakland's, they were right. At the same time, this meant that the federal government created alternative local infrastructures across the country, which local (and state) elites took as a declaration of war. They had effective allies in the local and national press, as well, all the way up to Tom Wolfe, who wrote a viperous, rac-ist, and curiously impossible-to-corroborate account of the program's functioning in San Francisco in 1969, "Mau-Mauing the Flak Catchers." Wolfe's account is particularly interesting in light of recent scholarship, led by Martha Bailey at the University of Michigan, among others, which has shown that the War on Poverty as a whole was fairly well administered and did, in fact, reduce poverty, despite being messy, un-derfunded, and philosophically disjointed.

The War on Poverty also set out to build leadership capacity within poor communities, and it did exactly that, incubating a whole genera-tion of Black leaders across the bay, shaping their expectations of what government could or should do. To look back at the rolls of people involved is to see a who's who of future political leaders: from Hun-ters Point's Elouise Westbrook to future congressman Ron Dellums, who worked in Hunters Point, to Bobby Seale and Huey Newton, who worked at the North Oakland local office—along with many more lead-ers whose power remained local, such as Adam Rogers in San Fran-cisco and Paul Cobb in the East Bay. (Not to mention Ms. Margaret.)

Dellums defended the program eloquently in his memoirs, recall-ing it as a "magnificent idea" that "provided the substantive answers to the follow-on questions raised by the civil rights movement," most of which boiled down to a simple reality: in America, if you didn't have money, then you could not actually exercise the rights you'd won. "The right to buy or sell a home in any community, established by success-ful legal desegregation efforts, was meaningless without the economic wherewithal to do so," Dellums wrote. "Jobs and economic develop-ment programs made real the promise of home ownership."

But the War on Poverty programs alone could not solve Mar-garet's problems. She was alone with two kids. She had fallen out of the middle-class ranks that her parents had fought so hard to lift her into. Her mother was drinking a lot. One day Margaret's father went to check on her brothers in a girlfriend's car. Her mother took a meat cleaver to its windows in a rage. Unsurprisingly, her brothers began to get into trouble, too.

Though Margaret finished high school, she didn't anticipate moving forward with her education. But a welfare caseworker she had at Double Rock encouraged her to go to San Francisco City College. She took the advice. Without regular child care, she'd drag the kids up to her mother's place on Vernon Street, then take the bus all the way back down to City College. The experience made her start agitating with other women for a child-care center.

The radical currents swirled around her. The slogan BLACK POWER could be found spray-painted around the projects. Stimulated in part by the Community Action Programs, young men formed the groups Youth for Service and Black Men for Action. Margaret's brother Billie became affiliated with the Black Panthers in San Francisco.

It'd been more than twenty years since San Francisco's Black population grew up around the shipyards of World War II, long enough for people to know that it wasn't different enough out by the Pacific, no matter the mythology of the place. "I thought coming to California, it was going to be a better place. It has been," the social worker Orville Luster told a documentary film crew. "But I think that once you wake up one morning you look out and see the Pacific Ocean and you see there is no place else to go. I must take a stand now. This is what has happened to a lot of Negroes who have come to the West Coast."

For many white people who moved to San Francisco, it represented liberation from the strict orchestral movements of East Coast urban life. Bohemians reading up in North Beach. Hippies smoking weed and dropping acid in Golden Gate Park. Queer communities living out loud. These were people who would not conform to the expectations of '60s society, who had simply declared themselves different, rogue particles in the cosmic gravy, people who were free to have new destinies. They could gather by the hundreds in a park and have no political message at all.

But where Ms. Margaret lived wasn't San Francisco, but Hunters Point, USA, as residents called it, a place apart from the white mainstream. Despite the vast numbers of young Black people born in the postwar years, they were left out of the narrative of the baby boom.

These people did not drop out but tuned in to the anti-imperial frequencies resonating across the world. This was true not just in Black communities but across the many peoples who had come to settle in the Bay Area. This unique configuration of people—which has only deepened in recent decades—gave local radical politics an unusual cast. While many Black movements in other parts of the country were stridently nationalist, Bay Area Black radical politics was international in its orientation. It saw common cause with countries outside the USSR-USA binary of Cold War politics, as represented by the Bandung Conference in Indonesia that became shorthand for this kind of solidarity. They also found common cause outside the Black-white binary, building links to Asian and Latin American groups. As the Vietnam War escalated, aided by what was going on down on the Oakland docks, anti-imperialist politics fused with local concerns.

Margaret's most direct route into this world was through the man whose name she would eventually take. Her mother had met this guy almost fifteen years older than Margaret. But he had a good union job as part of ILWU Local 6, the part of the union that worked in the growing warehouses in Oakland. "So my mother, in her alcoholic haze, introduced me to Benjamin Gordon," Margaret said.

His parents came from two venerable Black traditions: through his father, the Communist Party, and through his mother, the club-woman Talented Tenth. "This man, Gordon, was brilliant. Motherfucker had a brain. It was nothing for him to read three books in a day. And absorb it and articulate it," Margaret said. "But Gordon did not like little kids." Still, Margaret felt like she had to make compromises. Gordon encouraged her to keep going to school, which is how she ended up transferring to San Francisco State in the fall of 1969, after the massive, disruptive, and influential student strike there.

Across the world, in parallel and in conversation with the African American–led movement for civil rights, freedom fighters and nationalists across the global south had tossed out European imperial forces. To people from Accra to Delhi to Jakarta to Manila, it was not at all clear that they should subscribe to American-centered capitalism. Not only were there the decades of exploitation that imperial capitalist countries had visited upon them but within the United States itself, there was a clear social hierarchy in which people who looked like themselves— non-Anglo Americans—were at the bottom.

Inside the United States, young people began to apply that same analysis to white society. If most white Americans ignored the way the government was attempting to influence and, if necessary, suppress the freedom of people across the globe in the name of anti-communism and an ever stronger Pacific Circuit, Black people inside the center of the empire could not.

To many young activists, the analysis was obvious: Black America was an internal colony of the white mother country. In the Bay Area, these ideas began flowing through a network of young activists centered at Merritt College, which had the largest percentage of Black students of any white-dominant institute of higher learning in the country. Shortly after the Hunters Point uprising, Huey Newton and Bobby Seale formed the Black Panthers at Merritt, borrowing a popular icon for Black resistance at the time.

Seale and Newton had read Frantz Fanon's *The Wretched of the Earth* over and over, underlining passages. Here was this French Afro-Caribbean psychiatrist halfway across the world able to articulate the very problems that they faced in America. "Abandoning the country-side and its insoluble problems of demography, the landless peasants, now a lumpenproletariat," Fanon wrote, "are driven into the towns, crammed into shanty towns and endeavor to infiltrate the ports and cities, the creations of colonial domination." Who were the brothers on the block if not this lumpenproletariat?

Huey Newton was the designated intellectual of the operation, and the man who penned the most important pieces of the group's theory. Seale was the people person, the talker, the guy who could move the crowd. In the early days, they formed a potent combination of intelligence, courage, and street wisdom.

The Oakland that birthed the Black Panther Party for Self-Defense killed Black people. White residents were angry that their nearly all-white city had been settled by Black migrants from the South, hurting their property values and besmirching the reputation of the city. Politically, the city was deeply conservative, controlled by a small coterie of powerful Republicans.

For Black people, violence was an everyday part of life. Redevelopment authorities bulldozed homes and businesses. Numerous industrial facilities polluted the air and water and earth. Slumlords made money by letting their properties deteriorate, allowing the decaying properties to leach lead into the blood of children. Rats attacked poor people, while the government spent billions on the Apollo program to send white heroes to space. As the Gil Scott-Heron poem about the moondoggle goes: "A rat done bit my sister Nell / With whitey on the moon."

Kids ran the streets for lack of parks. Those same kids got hit by cars, and those accidents were then counted against the neighborhoods in the city's blight calculations. Pollution poured into the air and earth of Black neighborhoods, by design. Other factories abandoned the city, taking their jobs and money and leaving the environmental devastation they'd wrought.

Many people in the ghetto did not have enough to eat, nor could they afford health care, nor were there even medical facilities in poor areas. Black people were paid the least and still had the highest rates of unemployment, barred from most unions through bylaw or tradition, and if they got in, kept out of union leadership and the best positions. Meanwhile, the draft was taking young Black men and sending them off to fight in Southeast Asia. And despite all this, en masse, white people seemed to be mad about even the meager funding that had gone to the War on Poverty, especially when poor people had the audacity to suggest that the legislation's suggestion of "maximum feasible participation" meant that people in places like West Oakland should actually control programs.

And if any Black person stepped outside the narrow confines of life that Oakland allowed, the Oakland Police Department—97 percent white, mostly suburban, notoriously violent, and completely immune from complaint—stepped in. A police officer could almost certainly hassle, beat, shoot, or kill a Black person and face no consequence beyond the conscience. Black Oakland residents had been documenting police violence for decades *by the 1960s*.

The idea that white America had created ghettos for the same reason that the Nazis created ghettos was not beyond the pale of thought. As one young man in Hunters Point put it, "We would rather die fighting for our people than wait for a slow death." As later research would show, these neighborhoods *did* kill people slowly, shaving ten, even fifteen years off their lives relative to people in white neighborhoods.

During the summer of 1966, a flyer circulated in Oakland that the Black paper in Hunters Point, *The Spokesman*, republished. "Black people are the children of violence," it read. "We were brought to this country in violence during the slave trade, and we have been enslaved by violence since we've been here."

Two decades of postwar civil rights organizing hadn't gotten the Black residents of Oakland any political power. They were colonized, and they wanted to be free, by any means necessary.

"Black people like Leonard Deadwyler are openly murdered all over the nation, and their murderers are excused for 'accidental homicide,'" the flyer said. Deadwyler had been shot by a Los Angeles cop while frantically trying to get his wife, who was in labor, to the hospital. In modern times, it would recall Philando Castile, another Black man who'd done nothing yet was shot in his own car's driver's seat. Deadwyler's last words reportedly were, "But she's having a baby."

A young Thomas Pynchon covered the killing for *The New York Times*, rendering the seething anger of Watts, a year after the biggest, bloodiest uprising in American history. "The killing of Leonard Deadwyler has once again brought it all into sharp focus," Pynchon wrote that summer, "brought back longstanding pain, reminded everybody of how very often the cop does approach you with his revolver ready, so that nothing he does with it can then really be accidental; of how, especially, at night, everything can suddenly reduce to a matter of reflexes: your life trembling in the crook of a cop's finger because it is dark, and Watts, and the history of this place and these times makes it impossible for the cop to come on any different, or for you to hate him any less."

As in Watts, so it was in the flatlands of Oakland. "It is time we fulfilled our destiny of violence and returned the violence to Whitey's head!" the flyer went on. "It is no longer a case of militancy—we are fighting for our very lives!!!"

The "destiny of violence" echoes Fanon's *Wretched of the Earth*, the ur-text of the political militants who crossed paths in the East Bay in and around the Panthers. Fanon's accounting of Algeria's war for independence from France saw violence not only as inevitable but "cleansing." "The colonized masses intuitively believe that their liberation must be achieved and can only be achieved by force," Fanon wrote. The logical answer to these troubles was to challenge the colonial authorities.

Bobby Seale lived with his parents and young wife on Fifty-seventh, a half block from Grove Street and, across its wide expanse, Merritt College. A few years later, elevated BART trains would roll down the center of Grove. Two decades on, it would become another Martin Luther King Jr. Way.

But in the late 1960s, Grove was a central strip of Black Oakland. Black businesses clustered along the arteries—San Pablo, Grove, Shattuck. There were charm shops for a lady's wig needs, and barbershops where young men could get a process, a hair-straightening technique. Bars and underground speakeasies gave the Black residents in the modest neighborhoods between the big streets somewhere to go. Seale and Newton's North Oakland was not the ghetto but a borderland between the dense West Oakland poverty that faced the industrial bay and the Black middle class *and* student radicals of south Berkeley.

Seale's life had two poles. One was Merritt College, where he was an engineering student and a seeker of Black history and political education. The other was the North Oakland Neighborhood Anti-Poverty Center, where he worked in different capacities. Connecting the two were a couple blocks of Grove and another thoroughfare, Fifty-fifth. It was along this route that he would encounter Donald Warden, a fiery Black capitalist, who co-founded the Afro-American Association with fellow Berkeley law students.

The AAA brought together all kinds of young Black people into a concentrated program of Black nationalist study and outreach. This was the group through which Kamala Harris's parents met, and it attracted Seale's friend Huey Newton, who was hanging around many groups of the time. The Revolutionary Action Movement had deep links to international movements, and Newton and Seale joined its front group, the Soul Students Advisory Council at Merritt. As time went on, they began to diverge from the RAM line. They wanted more direct action.

Their first innovation was their police patrols, in which they followed cops in their own cars while heavily armed. That tactic faltered when state legislators passed a bill outlawing openly carrying weapons, which the Panthers protested by going to Sacramento and, as history has revealed, accidentally walking into the chambers. The shocking image of gun-toting (Black) men in the capitol made international news.

The Panthers were still mostly disorganized. Many Panthers were in their teens. That they were able to capture so much global attention and use it so effectively remains an incredible achievement. In their remarkable book *Black Against Empire*, Joshua Bloom and Waldo E. Martin, Jr., argue that the Panthers' anti-imperialist politics had such influence because the Vietnam War created the conditions for Americans to see the nation's power in new ways. But it took an event that occurred in the wee hours of October 28, 1967, on Seventh Street in West Oakland, literally right outside what is now Ms. Margaret's building, to bring the international spotlight to them.

Newton had ended up there after a Friday night touring what the Black Bay Area had to offer: dinner at home on Telegraph Avenue with his girlfriend, a drink at Bosn's Locker on Shattuck, church social down on Forty-second, house party on San Pablo, and a 4 a.m. trip to Seventh Street for late-night soul food.

No one knows precisely what happened, though many claim to, but afterward, one police officer was dead and Huey Newton was headed to jail. He became a cause célèbre, catapulting the Panthers from a small local group into a true international phenomenon. Suddenly everyone knew the Black Panthers. They became the symbol of anti-imperialism and Black Power fused together. Different chapters sprang up around the nation, and though there were some similarities, especially in the menu of social services that the Panthers tried to provide, each chapter was distinct and some were quite different.

One of their fastest-rising early recruits, George Murray, was teaching at San Francisco State. He rocketed from member of the Black Student Union at State to the Panthers' minister of education in the months after Newton ended up in jail. He was one of the key messengers for the Panthers' "Free Huey!" campaign along with Seale and Elaine Brown. On a trip to Cuba on behalf of the Panthers to attend the Organization of Solidarity with the People of Asia, Africa and Latin America (OSPAAAL) conference, Murray made a fiery speech. "The Black Panther Party recognizes the critical position of black people in the United States," he said. "We recognize that we are a colony within the imperialist domains of North America and that it is the historic duty of black people in the United States to bring about the complete, absolute and unconditional end of racism and neocolonialism by smashing, shattering and destroying the imperialist domains of North America."

He emphasized that the Panthers were not just against the Vietnam War, they were *for* the North Vietnamese, who were, in essence, killing the same people who oppressed Black Americans.

Back at SF State, the school moved against Murray, albeit slowly. Student resistance against his dismissal built. With support from the Black Panthers, the Black Student Union began to take more radical action. It sparked the creation of the Third World Liberation Front, which united groups representing what we would now call people of color. "The student strike at SF State and most of the critical campus rebellions that followed linked Black Power with a cross-race anti-imperialist perspective, often explicitly linking the fight for Black Power on campus to the Vietnam War and global anti-imperialism," write Bloom and Martin in *Black Against Empire*.

This was a time when it seemed history was moving fast. A site dedicated to memorializing the political rise of the Asian American Movement tried to capture the rhythm of the era. "You had to be there, sensing the world turning upside down. It wasn't remote or academic at all. On our TVs and in our newspapers we witnessed Asian faces rising up to finish off the latest colonial occupation," the site declared. "An entire quarter of humanity, once dismissed as clinging to a colorful past . . . had now stood up, an enormous Red banner of self-determination . . . There was no irony in a militant Black Power salute or a gentle wave of 'Peace, man.' *It was* real."

The Panthers were there to provide an alternative analysis to heading over to the jungles of Southeast Asia. Ericka Huggins, who both wrote for the Panther paper and ran a Panther school, noted, "The Black Panther Party was never nationalist, it was never insular, and if you look at the *Black Panther Intercommunal News Service,* you'll see . . . many articles about what was going on in Vietnam, in China, all over Africa . . . in Latin America, too."

Murray wrote an editorial about the SF State standoff in *Rolling Stone* that's still stirring. "Freedom is a state not limited to a particular culture, race or people," he stated, "and therefore, the principles upon which a struggle for human rights is based must be all inclusive, must apply equally for all people."

The Third World Liberation Front groups all found different ways into common cause with Black students. For the Philippine American Collegiate Endeavor, the call to struggle could be seen not just in episodes of anti-Black racism but in the "small number of Filipinos in college, opportunities denied to Filipino professionals in this country, [and] exploitation of Filipino farm workers in Delano working for a few dollars a day." For the Intercollegiate Chinese for Social Action, San Francisco did "not begin to serve the 300,000 non-white people who live in this urban community in poverty, in ignorance and in despair. The Chinese ghetto, Chinatown, is a case in point." For the Mexican American Student Confederation, the excitement lay in connecting on-campus student protests with "our growing ties to Chicano and Black working people." They promised to "wage an uncompromising struggle for Third World liberation."

It was these groups that joined the Black students in one of the most disruptive student actions in this country's history. The school was shut down over and over. The Big 5 from Hunters Point came across town to support the strikers, the historian Aliyah Dunn-Salahuddin pointed out. "I want you to know that I am a Black woman, I'm a mother, and I have fifteen grandchildren, and I want a college that I can be proud of," Elouise Westbrook told a crowd at the school in December 1968.

A new president was appointed, S. I. Hayakawa, and he liberally used police force to try to restore the status quo. Nonetheless, the strike dragged on. The Third World Liberation Front spread to other campuses, as did the strike tactics. The struggle helped to solder together single ethnic groups into larger configurations. While Latin American (not Mexican, Guatemalan, Nicaraguan) was a designation with some history, Asian American (not Chinese, Japanese, Filipino, Korean, Vietnamese) was not. A group of Berkeley students declared themselves the Asian American Political Alliance. While they would struggle with what material and cultural conditions a third-generation Japanese American college kid might share with a new arrival from Vietnam or a Cantonese-speaking longtime resident of Chinatown, Third World solidarity along the model of Black Power provided a powerful organizing idea. At Berkeley, the Third World Liberation Front even more or less lifted its constitution from the Afro-American Student Union, crossing out "blacks" and writing in "Third World people."

In the already hyperdiverse Bay Area, it was the white power structure versus the Third World. So many current Bay Area Asian leaders launched out of that time. Jean Quan would become the first Asian woman to lead a major American city. She gave fiery speeches on behalf of the Third World Liberation Front to colleges around the area.

A talk she gave at Cal State Hayward in the East Bay was enough to inspire Pam Tau Lee to join the radical movement. Later, Lee would go on to help found the Asian Pacific Environmental Network, an important environmental justice group in Richmond, precisely because the internationalist energy around EJ reminded her of the Third World days.

Perhaps the SF State strike was not a total win for the strikers. Not all their dreams of control over the curriculum and professoriate were realized. Murray, who was among the hundreds jailed, was browbeaten out of politics. But the strikers won an ethnic studies program that would be copied across the nation, and they inspired other students with their direct-action tactics.

Locally, they massively increased enrollment for Black and Third World students, one of their key demands. And that's the longer story of how Margaret Gordon ended up transferring to San Francisco State just a quarter into her college career. She'd moved in with Gordon, and with his full-throated support, she'd trundle over from Oakland, leave her kids with her mother, then go to school.

The classes she took reflect both the standard student experience (English Freshman Seminar—Pass) and the influence of the Third World movement. She got an A in her course "The Ghetto and the Slum." Margaret had gotten out of the projects and was on campus at one of the most exciting schools in the country, in part thanks to the Third World Liberation Front and the Black Panthers, and in part thanks to Gordon. She was grateful.

Mike Vawter was a working-class white kid in the Bay, born in the same Hunters Point where Margaret Gordon lived, back when it was a "mixed-race Navy ghetto," as he described it. His family moved out to Pacifica, and he rode his natural math abilities all the way into Stanford, where he got radicalized by the anti–Vietnam War movement.

Many Stanford activists were leftists in the Graduate Coordinating Committee, and they began the work the way a scholar might: creating a map of the connections among the Stanford Research Institute, the Hoover Institution, Stanford proper, the trustees of the university, the military, the intelligence services, and the corporate suppliers of the armed forces.

Because the links between these institutions had been assiduously cultivated by elite administrators such as Provost Fred Terman and President Wallace Sterling, there were many, many, many connections between Stanford and the Cold War apparatus. Money flowed freely from the war machine to the university. David Packard alone was evidence that Stanford had become central to the postwar economy. He was chairman of the board at HP, of course. But he was also a director at Pacific Gas & Electric, U.S. Steel, General Dynamics, Crocker-Citizens

Bank, and National Airlines. Other trustees were executives from Standard Oil, Union Oil, Shell, Lockheed, General Dynamics, mining concerns, the chemical company FMC, and RAND.

To emerging student radicals, the interrelationships of the network were obvious and deplorable. What and who was the ruling class? It wasn't abstract—it was the people who ran the university and their friends. David Ransom, then an instructor in the English department, called it the "Stanford Complex." SRI carried out the technical dirty work of learning how to fight America's imperial, anti-Communist wars. The Hoover Institution grounded this work in ideology that guaranteed increased military spending, which fed back into SRI and the Hoover Institution itself. The engineering department would push new graduates into the surrounding corporations, and the corporations would supply the latest technologies back to the school. Each piece of the university had its place at the feeding trough of global, never-ending imperial war.

"That this complex should be deeply involved in all aspects of the Vietnam War—military, economic, political—should come as no surprise," Ransom wrote in 1967. At its center lay SRI. Its analysts had helped to manufacture the idea that Southeast Asia was a crucial strategic link, one arguing that "the free world must not lose Southeast Asia." SRI employees had also done research on chemical weapons and the industrial partnerships across the Pacific. By the late 1960s, SRI had teams deployed in Thailand to research "counter insurgency."

Ransom lambasted Terman for describing his "community of technical scholars" as the "wave of the future." To Ransom, it was "a military-industrial-university complex at war with half the world."

Stanford's trustees and institutions *were*, of course, part of the war machine. That had been the whole point of Stanford to the World War II men. They supported the military and the military supported them. They were a group of people whose interests, backgrounds, and intuitions about geopolitics were mostly aligned.

But students like Vawter—and their young teachers, like Bruce Franklin, who had served in the Air Force—were increasingly unsure of what to make of the congealing of so much power into an interlocking structure of old white men. Vawter spent the summer of 1967, that famous Summer of Love, living in the Haight with a girlfriend. Peace and love and sex were great. But when he returned to Stanford that year, he ran into Franklin, who had begun taking a more radical line, promoting a revolutionary change in the U.S. government. Students like Vawter began to get very serious. "What made things move so fast at Stanford was the nature of Stanford itself and its interrelations not just with Silicon Valley, but the empire and the war," Franklin told me.

By 1968, Vawter had joined the Red Guard, as part of a coalition of Maoist groups co-led by Franklin. Through the next year, the anti-war activists put pressure on the school to sever ties to SRI. They held sit-ins, teach-ins, and protests. They circulated publications and recruited new people to the cause. Students for a Democratic Society served as the visible face of the student radicals.

In March 1969, the trustees agreed to an open forum with students, presumably to talk some sense into them about the SRI issue. Bill Hewlett had been on the board of FMC, a military contractor, which students had accused of making nerve gas. Hewlett had denied it for a year, as Vawter recalled. Then, when a student put it to him that FMC's production had been confirmed by the journalist Seymour Hersh and the students' own digging, Hewlett denied it personally in front of the crowd of thousands at Memorial Auditorium. "I'm amazed by the accuracy and reliability of your sources," Hewlett said, "but I happened to check with the president of FMC, whom I consider superior to your sources, and he says that they are *not* making nerve gas at the present time."

At which point, the student made the obvious follow-up: "Have they *ever* made nerve gas?" Hewlett had to admit that the company had been "asked to build a plant, which they built and operated at the request of the government and they turned that plant over to the government about six months ago."

The crowd was stunned. They'd been lied to all along. "The spontaneous collectively shocked gasp of thousands was audible," Vawter said. It was followed by, as a transcript notes, "loud laughter and applause."

After the meeting, an umbrella group known as the April Third Movement formed to ramp up pressure on campus, occupying the Applied Electronics Laboratory and Encina Hall. As at SF State, the student activism did not win totally outright. These were powerful institutions with a lot of other support.

Eventually, Stanford did spin SRI out and agree to stop doing chemical and biological weapons research. This was significant. The uncomplicated link between Stanford and military funding had been broken. It *meant something* to take money from the military, and it meant something to refuse it.

There were real costs to the victory. Vawter was thrown out of school for his role in a confrontation and went down to the waterfront to work for the ILWU. Bruce Franklin, Vawter's mentor, was fired from Stanford, still the only tenured professor to receive the honor.

He and his followers threw in with the revolutionary Chicano group Venceremos, founded by Mario Manganiello. They primarily operated out of East Palo Alto, pressing Black Panther–style tactics in the Peninsula. They held street parties, operated a food co-op, supported local unions, created abundant media, *and* advocated direct armed resistance. One of their favorite targets was David Packard, whom Richard Nixon had appointed to head the Department of Defense. The radicals could not have made a better case that HP's products were essential to the prosecution of the war.

As representatives of the local power structure as well as the American empire, Hewlett, Packard, and their company fused the levels of injustice that Venceremos saw. The community's problems could not be disentangled from the global ones. The way Venceremos saw it, Hewlett Packard was a major supplier to both the military *and* the rest of the military suppliers. They were a crucial part of Stanford's electronics cluster, and they were "molding the midpeninsula to suit their class interests" by putting the research labs in the nice cities with nice housing and the assembly plants in poorer areas. "Thus Stanford has divided the Peninsula neatly along class lines," they wrote.

HP wanted the war to be one of those things about which reasonable people can disagree. The radicals did not. Some members of

Venceremos were willing to do anything to stop the war. Eventually four were convicted of murdering two corrections officers. As the war nose-dived toward a close and under severe pressure from law enforcement, Venceremos could not sustain itself and folded. They had not *won*, but they'd shown how the Bay Area had become the means by which Washington, D.C.'s imperial orders were carried out.

The student radicals and their allies had created a theory for the university's role in geopolitics. The American empire was securing labor and resources abroad through the application of information technologies that allowed an ever smaller number of elites to control ever more people. Stanford was the center of the effort to extend that control. So the fight was not really over the university, or even over a factory in a town in a state in a nation, but a network of many, many factories spread over the entire Third World.

Margaret spent only a year at SF State. For all the students' excitement about building bridges to the ghettos of the Bay, neither the activists nor the traditional school infrastructure devoted a lot of attention to what supports a mother like her might need. Was there mental health or relationship counseling? Child care? Transportation help? Many activists were middle-class, regardless of ethnicity, and very few were the primary caretakers for children.

So the commute took its toll. Always dragging the kids took its toll. She had discovered that she could not leave them home with Benjamin Gordon. She transferred over to Merritt College, the original home of the Panthers, and devoted her activist energy there to getting child care set up. They won that battle.

But life was not good. Later, she could look back and could see how he traumatized her and the kids, but at the time, she clung to the lifeline that he had thrown her.

Her family tried to take care of her. The kids began to bounce for longer periods between Margaret and her mother, but Clemmie's descent into alcoholism had accelerated.

Margaret was no stranger to domestic violence. She'd gone toe-to-toe with her first husband. With Gordon, it was different. The details remain within the family. All Margaret's son Tony would tell me was, "I don't got nothing nice to say about Gordon."

The times, too, became deflating. The Panthers had overextended and were weakened by FBI counterintelligence, Newton's own drug use, and all kinds of violence inside the organization. As the Vietnam War drawdown got going, the support for the Panthers' more radical agenda began to dry up. Mass student protest energy waned. Some began to focus on organizing in their communities, such as the Asian American–led urban renewal battle over the International Hotel in San Francisco.

The world had seemed to be moving *so fast*, and then it slowed down. Not everything changed. The revolution did not come. Things were stuck. Margaret herself was locked in exactly such a condition. In 1974, her mother, Clemmie, died. Auntie Emmer—the one from down in Arkansas, then Richmond—had long since moved to Alaska. When she came down to the Bay for Clemmie's funeral, she saw Margaret mired in this terrible relationship. So Emmer took Margaret's kids back to Alaska, acting as a kind of foster parent while Margaret righted the ship.

Tony, ten, and his brother, Jarrold, eight, went up north to Fairbanks. It was cold up there. Beyond cold. There are sweet pictures of them from the time, entirely '70s photos with houseplants and odd fabric patterns. In one, Emmer sits at a table, probably around Christmas, piled high with food, a bottle of champagne popped open. She sits upright with her hands in her lap. A friend is laughing. George, her husband, stands in his bathrobe in the background, cigarette in mouth, coffee on the stove.

It wasn't perfect or fancy, but Emmer had made a life in Fairbanks. She worked nights for the state of Alaska as a janitor. She'd built a home, adding chunks to it as she taught herself carpentry. "She was like a sponge for knowledge," Tony remembered. George took care of most of the domestic stuff. He drank too much when his paycheck came in, but he and Emmer were good to the kids. They did their best to teach them how to live.

Emmer was a complex figure. She had not experienced a real childhood, and never learned to really read. She didn't like to talk to white people. They might try to make conversation, and Emmer would say nothing. "She would chew tobacco and not talk, her mouth full of spit. She'd be like, 'Mmmhmm,'" Tony remembered. Then she'd go over to the garbage can and spit and tell him, "'I didn't want to talk with them, Tony. I didn't want to talk with them.'" But she could pick up anything by watching and doing. "Just imagine if she had a college education and a degree and everything," he said. "You move from Arkansas to California, California to Alaska, you have no limits."

No matter how far she went, farther and farther and farther from where she was born, some of the lessons of Crossett and Jim Crow never left her, as Tony discovered. During the winters, there was almost nothing to do but play sports, so he was on the basketball team. A teammate started giving him shit. "Kid kept picking on me, picking on me. It was a white kid. So I turned around and punched the kid in the nose and he started bleeding," Tony said. "Emmer just went crazy. She went and found a switch and beat me with it. 'Don't you ever do that to a white person. Do you know what they could do to you?!' They all knew I was not in the wrong."

But that was the whole point: it didn't matter if he was right or wrong.

13

The Technology Question

Is it possible to imagine U.S. agriculture and service industries
without Mexican migrant labor, or Arab oil without Palestinians
and Pakistanis? Moreover, where would the great innovative
sectors of immaterial production, from design to fashion, and from
electronics to science in Europe, the United States, and Asia, be
without the "illegal labor" of the great masses, mobilized toward
the radiant horizons of capitalist wealth and freedom? Mass
migrations have become necessary for production.
—MICHAEL HARDT AND ANTONIO NEGRI

In the 1970s, the Panthers retrenched to Oakland, calling in members they deemed loyal from across the country. Newton remained mired in legal battles, and Elaine Brown climbed upward in the party hierarchy. Their tactics continued to evolve. They expanded their "Survival Pending Revolution" programs, providing food, education, and health services. They would expand their traditional political presence and eventually take over some pieces of Oakland nightlife.

It was a complex operation. At one point, the Panthers owned eleven houses through their corporation, Stronghold Consolidated Productions, Inc., including six in Oakland, as well as two in Berkeley and one each in New York, New Haven, and Chicago. They also maintained a penthouse "office," at least as noted in the accounting documents created by their bank, Wells Fargo, in a tall apartment building at 1200 Lakeshore, on the inner east side of Lake Merritt.

As Newton's long legal battles played out, he lived in that penthouse, and from there he could see out over the city and right into the growing Port of Oakland. The cranes, the containers, the new Japanese ships. This was, Newton realized, the working mechanism of imperialism by the 1970s. Richard Nixon's Guam Doctrine indicated that the United States was not going to fight every war in every place, Cold War or no. Instead, the military would create a "nuclear umbrella," and then bind every country in Asia ever closer to the United States with economic power.

Each box loaded and unloaded tightened the links between Babylon, as the Panthers called America, and the rest of the world. Between the burst in trade and travel, alongside the growth of mass media, technological developments had created an "undeniable interconnection to everything among all the territories of the world," Newton wrote. And right then, right there, Oakland was the most important link between the United States and the emerging economic powers of Asia.

I struggled with the valorization of Huey Newton for the whole time I was writing this book. While his accomplishments are truly remarkable, the evidence that he became violent and abusive is solid and well documented, by those inside and outside the BPP.

Nonetheless, Newton's position as an anti-imperialist radical thinker in exactly the place where the global economy was undergoing such tremendous change was a remarkable intellectual opportunity. Newton developed a new theory of what we would now call globalization, even as he began to descend into disordered drug use and more general running of the streets.

Newton became keen to the ways corporate power was simply *not American*. It might have American roots, but it was something all its own. "In order to plan a real intercommunal economy we will have to acknowledge how the world is hooked up," Newton said. "We will also have to acknowledge that nations have not existed for some time." Global capitalism had eviscerated the power and primacy of nations. "Their self-determination, economic determination, and cultural determination has been transformed by the imperialists and the ruling circle," Newton wrote in his essay "The Technology Question."

At the same time, technology had made new forms of power possible. Newton theorized that American power players did not want a traditional colonial empire, occupying territories. They didn't care about "the land question," as he put it, but rather the markets represented by all the world's people. They wanted consumers and laborers. "The people of the oppressed territories might fight on the land question and die over the land question," Newton wrote. "But for the United States, it is the technology question, and the consumption of the goods that the technology produces!"

Huey Newton began to extend Fanon's vision. New technologies were changing the conditions of the lumpenproletariat. Ruling elites had scooped up the world's resources through imperialism, both the old European ones and the emergent American version. But the wealth was not simply made into gold bars or turned into hundred-dollar bills that were lit on fire. That capital allowed the wealthy countries to build institutions (like, say, Stanford) that created "the technological machine."

The United States didn't need to occupy and run Korea or Malaysia as colonies. Instead, it could direct the development of the country from afar, so that it would provide the labor and consumers that the U.S. sought. Newton had a crucial realization about supply chains: because they dispersed the ethical responsibility for the violence inherent in the system, no one even had to buy into the military-industrial complex's aims. "The U.S. capitalist has been able to spread out his entire operation. You put together his machinery in parts," he wrote, "thus you are not building a bomb, you are building a transistor."

Those transistors, packed onto silicon semiconductors and packaged into electronics, were remaking the economy and culture, too. "They raise the standard of living through transistors in order to further rip-off/sell its goods to the workers and the people of the world," Newton said.

The American system produced products that people liked, even loved. The U.S. could say, truthfully, "We're producing for everybody; we're giving out the goods." Newton told the story of Alex Haley visiting Africa. Haley had visited a village and saw "an old man walking down the road, holding something that he cherished to his ear. It was a small transistor radio that was zeroed in on the British broadcasting network." The man had a product containing Silicon Valley's R&D output playing the mass media of the empire that had colonized his country.

There *was* something legitimately democratizing about this new system.

The technological system meant that nice things could cost less. You get a radio and you get a radio and you get a radio. Then it's a TV. A computer. A cell phone. And if the components of these technologies were also the backbone of the U.S. military, and if the vast sums of money being made selling them were accruing to a white, male elite in California, well . . . you still wanted your TV. And then your color TV, then your HD TV, then your 4K TV. Your iPhone 14 and iPhone 15 and iPhone 16. "The technological question is unopposed—as far as who benefits from it, because we all do on one level or another," Newton wrote.

Newton quoted a Ford executive on their operations: "It is our goal to be in every single country there is. We look at a world without any boundary lines. We don't consider ourselves basically American. We are multi-national; and when we approach a government that doesn't like the United States, we always say, 'Who do you like; Britain, Germany? We carry a lot of flags.'" Newton realized that what we ended up calling globalization was accelerating. This economic arrangement was what the Panthers were up against, not just the U.S. state apparatus or the mayor of Oakland.

The nature of the Pacific Circuit meant that many different peoples were finding themselves in the same position relative to the economic system. "We see very little difference in what happens to a community here in North America and what happens to a community in Vietnam. We see very little difference in what happens, even culturally, to a Chinese community in San Francisco and a Chinese community in Hong Kong," he said.

All these people had lost some key rights that went along with nationhood, including the ability to get some purchase on power. Perhaps more people were nominally free, but they were still subject to the economic and cultural dominance of mostly white elites who used the control of capital, territorial segregation, and (if necessary) violence to maintain power.

Tech, in its various manifestations, allowed the ruling elite to bypass traditional national structures. Seizing control of the land or the means of production in one country was fine, but it would not mean you truly ran the nation. After North Vietnam won the war, what would or could the peace look like without transnational capital? As many of the postcolonial countries across the world discovered, there were other forms of control that were extremely powerful.

A technologized, financialized global elite had created what Newton called "reactionary intercommunalism," which meant something like what we call corporate globalization. Newton accepted that intercommunalism would be the order of the day, no matter what, but the politics underlying it could be different.

Solidarity might be possible to build across wildly different cultures. They'd seen it themselves during the SF State strike, when the Panthers spoke right alongside the leaders of the Third World Liberation Front.

How, though, should they take on the system, or build a different one?

While Newton's apartment looked across and down to the port, the Panthers' stronghold in West Oakland directly abutted it. Some of their members worked as casuals, and while the evidence that the historian Peter Cole managed to dig up on ILWU-Panther overlap was surprisingly thin, the leaders of the Panthers became obsessed with the port. During the Panthers' turn to traditional electoral politics in the early 1970s, both Bobby Seale's mayoral campaign and Elaine Brown's city council run emphasized the port as the centerpiece of their economic program. In her memoir, *A Taste of Power*, Brown recounts that the Panthers knew the port was significant, and that it was part of a new form of economic arrangement, one that had already superseded the muscular nationalism of mid-century America.

"Oakland was a city in which the near-majority black population had something to seize. Oakland had a durable economic base, one that was tied immutably to the city government. It had the Port of Oakland. The capitalist club in Oakland had revolutionized the port with new technology. It was one of a handful in the world that were fully containerized," she wrote. "Oakland's port enjoyed a strategic location. It furnished access from the Pacific Ocean to all of America, and from all of America to the other side of the Pacific. As a result of the combination of containerization and natural position, the volume of Oakland's port business had increased phenomenally."

The Panthers wanted to take over the city, not just to run city hall but to run the port. "Within five years, Oakland could become a base for black liberation," Brown concluded. "Within ten years, Oakland could become a base for revolution." Oakland could become the first node in that new anti-capitalist network of revolutionary communism.

The technological machine would create new conditions, namely mass unemployment and immiseration via automation, opening up new people across the world to the Panthers' message. In this telling, ghetto residents were not a group that had been left behind but harbingers of runaway inequality. "While the lumpenproletarians are the minority and the proletarians are the majority, technology is developing at such a rapid rate that automation will progress to cybernation, and cybernation probably to technocracy," Newton said.

Bobby Seale, always a man of the people, put it in simpler terms. "The technology is so great and the automation is so perpetual in its scope and its development," Seale said, "that it creates these unemployables that must engage in other activities to survive!"

To the Panthers, ghetto residents were *the early adopters of the future economy*. And at least some of the conditions of the ghetto have come to ever broader swaths of America: dead-end jobs, informal work arrangements, indebtedness, despair-induced drug dependency, exclusion from power, predatory inclusion in financial systems, pathologization from above, permanent renterdom.

And that was just at home in Babylon.

←----   ----→

At 10:15 a.m., on November 26, 1978, on the Malaysian island of Penang, in the corner of an American-run factory, a young woman began screaming, believing her soul to be possessed by an evil spirit. The cries and shouts spread across the shop floor as more and more women battled the ghosts, known as *datuk*. In total, fifteen women were affected. As supervisors tried to restrain them, some violently resisted. Others simply sobbed. The affected workers were given sedatives. The remaining women were sent home as the factory shut down for three days. Workers said that a ghost was haunting the dirty toilet they had to use while on the production line.

These incidents happened regularly at production facilities on Penang, known regionally as "Silicon Island." Women sank into trances, went into seizure-like convulsions, or became unconsolably terrified.

"In one possession which I witnessed, ten adults were needed to restrain a very slight teen-aged girl," the reporter Rachael Grossman wrote in a joint issue of *Southeast Asia Chronicle* and *Pacific Research*. "In another, a worker who was possessed in her hostel began to shout that she hated being there, hated working in the plant, and wanted to go home to her mother. Afterwards, she and others went to great pains to explain that it was not she who was speaking but a spirit who was speaking through her." Plant managers hired psychologists as well as *bomoh*, local spirit healers, who might, for example, sacrifice a goat on the premises to appease the angry ghosts. One occupational psychologist published a paper titled "How to Handle Hysterical Factory Workers."

Grossman saw the episodes as "one of their few culturally acceptable forms of social protest" against the work, an interpretation expanded by the anthropologist Aihwa Ong to include the other dislocations of corporate work in what had been a rural society. "They are acts of rebellion, symbolizing what cannot be spoken directly, calling for a renegotiation of obligations between the management and workers," Ong wrote.

These women found themselves in the crucible of the new form of industrial production—the integrated global assembly line, or as it would come to be called, *the supply chain*. Labor activists across the world realized that the electronics industry was driving this new labor model, employing women across vast networks. To these activists, it was not a secret how this worked.

"In California's Silicon Valley, for example, research and development work is carried out by well paid, usually male scientists and engineers. Circuits are then photographically etched onto layers of silicon in nearby assembly plants by women—50% of whom are Asian or Latin—for low wages in a highly pressurized environment," wrote the British sociologist Diane Elson for a conference called Women Working Worldwide. "The new stage is relocated to Southeast Asia where the silicon slices are cut up, bonded onto circuit boards, sealed in ceramic coating and tested. From there components are sent to other third world countries to be assembled into watches, etc., or sent back to the U.S."

If one end of the supply chain was a lab in California, the other was a factory carved out of a tropical island, the dictates of Palo Alto directing the labor of Penang. The same celebrated Silicon Valley figures who created the casual, fun culture of HP, Fairchild, Apple, and all the rest *also* took advantage of the Pacific Circuit to access vast numbers of impoverished, repressed workers.

It might be nice to think of American companies as friendly partners to local businesses, and at a personal level, that might even have some truth to it. But on the global stage, American military power helped prop up "anti-Communist" authoritarian regimes across the region, which just so happened to create the kinds of labor and infrastructural conditions that American electronics and semiconductor firms could exploit. In Indonesia, it would be Suharto. In the Philippines, Marcos. In Korea, Rhee. All were encouraged to create export-driven economies, which tied their elites to the interests of American geopolitics, and many took on American military facilities, tying their security to the empire, too. American aid to Thailand, for example, took the form of "highways, airport facilities, and deepwater ports," the very things that would make global supply chains possible and efficient.

Low-wage countries around the world set up "free-trade zones" or "export zones," mostly in Asia. According to labor researchers in the early '80s, about half the workers in those zones were working for electronics companies, or what we'd now call tech. In one documentary, *The Global Assembly Line*, a woman whose house was in the Bataan Export Processing Zone in the Philippines described a process of neighborhood demolition with disturbing echoes of the urban renewal programs in the United States. The government simply sent in the military to uproot the entire fishing community and move it away from the shore, because it wanted the land along the Pacific for industrial development.

The mother of nearly all of Silicon Valley's important semiconductor firms—Fairchild—had begun outsourcing in 1963 in Hong Kong. In subsequent years, it expanded to Korea, Singapore, and Indonesia, the latter three all places where U.S.-backed authoritarians kept labor in line. Other companies followed suit. "By the mid-1970s, there were dozens of U.S.-owned assembly plants throughout the region employing about 1,000 workers each," one history of offshoring found.

Women in the Bay Area and women in Asia were, as the radical Pacific Studies Center noted, "on different sections of the same assembly line," yet labor unions could not see how to organize together in the way that the transnational configuration suggested. It helped the corporate powers that unions were often banned in free-trade zones.

Within the geopolitical game, high-technology companies had a special role to play. The garment industry would clearly not be the route forward as the rural masses were pushed and pulled into Asian cities. Many Asian economic planners saw electronics as the path to development, building expertise and infrastructure in the parts of the business that Americans did not or could not handle. American experts like Stanford's Fred Terman were integral to these efforts. For example, after Korean officials got millions of dollars of loans from the United States to build out the Korea Institute for Science and Technology,

Terman headed a committee to "assess the feasibility of KIST's proposals and projects."

If such economic progress meant violently repressing labor organizing, then it meant violently suppressing labor organizing, all while keeping the American companies a step away from the action. "Workers must uphold their dignity and not cause problems that would scare away foreign investors," the deputy prime minister of Malaysia wrote. "They should instead be more productive so that the government efforts to attract investors would be successful." Their advertisements to Western and Japanese companies made their case clear. "The manual dexterity of the oriental female is famous the world over," a Malaysian publication read. "Her hands are small and she works fast with extreme care. Who, therefore, could be better qualified by nature and inheritance to contribute to the efficiency of a bench-production line than the oriental girl."

An Intel manager even said the quiet part out loud: "We hire girls because they have less energy, are more disciplined, and are easier to control."

At the same time, American companies strove to create an image of a new kind of womanhood, a cultural imperialism that resulted in grotesqueries like the "Miss Free Trade Zone" pageant in Penang. Different foreign firms would have intra-company competitions, and then send their winners on to the big one. In one National Semiconductor newsletter, "Miss Maria" stands in a bathing suit and platform heels with the number 13 emblazoned on her outfit. She was a bonder, attaching pieces together. She would be National Semiconductor's entrant in the Miss FTZ Beauty Contest, where she was "hoping to bring back joy and glory to the No. 1 Company."

In Malaysia, the contrasts between rural Muslim village life and the electronics agglomerations in the free-trade zones were particularly stark. While Grossman reported that some women happily escaped the strictures of the village for the "freedom to go out late at night, to have a boyfriend, to wear blue jeans, high heels and make-up," the country's social infrastructure was not fully prepared for the changes in society's labor conditions. Women who worked in factories could be seen as damaged goods, even if their incomes helped to support their families in the outlying regions.

These tensions played right into the hands of managers. In some factories, the male bosses saw themselves as father figures to the women on the line, with all the attendant problems of power and patriarchy. In other cases, women were desperate to keep their jobs not only for the income but because their social status was tied closely to the factory work they were doing. If they lost a factory job after tossing off the village norms, there were few alternatives for making a living. Except, that is, another job category that was proliferating through Southeast Asia—sex work—often clustered around the very same foreigner-heavy enclaves of free-trade zones and American military bases.

Silicon Valley's success created huge fortunes and hundreds of thousands of new jobs not just in Asia but in Santa Clara Valley, many in manufacturing. The boom coincided with a change in American immigration policy, the Hart-Celler Immigration Act of 1965, which opened up entry into the United States from Asia and Latin America. As refugees fled the widening gyre of instability in Southeast Asia, many arrived in the United States and were put to work in Silicon Valley factories.

The sociologist Saskia Sassen also argued that the very kinds of investments and export-oriented development deals that the United States made across Asia formed "linkages that contributed, directly or indirectly, to emigration." With the disruption of traditional labor forms, introduction of more Western cultural products, and building of relationships between places, making the move to leave for America became more thinkable. It's another one of those feedback loops that make the Pacific Circuit such a transformative force.

Already by 1971, more Asians were immigrating to the United States than Europeans. Over the course of the 1970s, the number of Asian people increased by 100 percent. The Korean population went up by more than 400 percent. The burst of refugees who arrived after Saigon fell meant that an estimated 11,700 Vietnamese people lived in San Jose by 1980, a number that would explode over the following decades.

This new population had to try to scrape together a living in an expensive new place. Many immigrant women, Asian and Latin alike, entered the workforce. Silicon Valley hiring managers looked out for them specifically because they were seen as high-productivity workers. The San Francisco State historian Karen Hossfeld interviewed many women working in the plants, who related stories of discrimination and the simple difficulty of making life happen with kids and a job.

Domestic managers followed the same sexist rules as their counterparts abroad. "Just three things I look for in hiring: small, foreign, and female," a manager told her. "You find those three things and you're pretty much automatically guaranteed the right kind of work force. These little foreign gals are grateful to be hired—very, very grateful—no matter what."

Other researchers found hiring managers describing the ideal worker as "the 'FFM'—or 'fast fingered Malaysian.'"

With attitudes like that within Silicon Valley companies, it is no surprise that in the manufacturing heyday of the semiconductor industry, Asian and Latin women were clustered at the bottom of the economic hierarchy, doing only the most manually intensive jobs. Labor organizers found that companies liked to splinter off the different ethnic groups. "Company A would hire almost entirely southeast Asians. Company B would hire mainly Latinos," one activist recalled. "You had segregation between corporations and facilities, by occupation, by shift, by job assignment, right across the board." By and large, they hired very few Black workers.

Many of the worst jobs were also dangerous, with regular exposure to toxic chemicals, Many, many workers' experiences are documented in the book *The Silicon Valley of Dreams,* by David Naguib Pellow and Lisa Sun-Hee Park. Some of the most harrowing stories actually occurred outside the clean rooms. Printed circuit board assembly became a popular kind of piecework. Vietnamese families would solder together, parents and children alike inhaling the lead and other chemicals with no safety protections. They did what it took to survive, just as they had back home.

←---   ---→

In the way that autoworkers made Detroit great, Asian women literally built the technology industry's products. This intersectional part of the story has been massively undersold, at least since the 1980s, despite the occasional flare of interest around Foxconn.

I do want to acknowledge the many facets of Silicon Valley, and the different stories we could tell about what created the fountain of wealth that would eventually overrun the whole Bay Area. Each contains important truths.

The technologies that generated the vast fortunes of the valley nearly all trace their roots back to military funding, needs, or inspiration. The military bought substantially all Fairchild's early integrated circuits, allowing the company to scale up, driving down costs, pushing chips onto the curve of Moore's Law, and spawning the Fairchildren. Missile programs drove the miniaturization of components. Anti-missile programs created new kinds of networks and interfaces. Supercomputers, as they were known, were driven primarily by the needs of nuclear weapons researchers. We saw how the Korean War laid the foundation for Silicon Valley. And that's not even talking about the internet, which was, of course, funded by the military from the mid-1960s through the 1980s. The historian Margaret O'Mara laid out the most sweeping history in her book *The Code*.

You might also look at how American bureaucracies needed computational and logistical technologies. To run social security, the government needed to be able to sort and collate cards, which is to say people and employers, at scale. IBM built a suite of machines that allowed SSA officials to operate a mechanical database of 26 million people, nearly a decade before the first general-purpose electronic computer. Queries could be run by pushing cards through sorters, which could split out, say, everyone in the Pacific states. With each update from employers, employee contributions could be tallied, then custom-designed collating machines could place the cards back in order.

This kind of punch-card system got its first large-scale trial during the 1890 Census. Initially created by the statistician Herman Hollerith, punch cards turned human information into machine-readable data. After many twists and turns, Hollerith's company became a core component of International Business Machines. IBM, along with its competitor Remington Rand, created an ecosystem of machines and processes to perform precisely the operations a large organization would need. And shimmering inside the mountain of operational data flowing through these stacks of cards was this byproduct, surplus information, that if captured correctly became valuable in itself.

Computers, improving rapidly thanks to the semiconductor industry, made this work much easier. The era of the mainframe began with banks and the military and other huge institutions. And then, like rats overrunning a ship, the weirdos and nerds of California began to use any unused cycles to create a different kind of computing, leading eventually to the home-brewed PC, Apple, and all the rest.

To understand what became "tech," you could look more theoretically, too, at the development of the very concept of information. This is a sympathetic story, coursing with the genius of Claude Shannon and the early creators of the transistor, all of them toiling away at Bell Labs, that great positive externality of the Bell Telephone System's monopoly on American telephony. Shannon took what could be stored in transistors and placed it very close to the center of existence, on the same plane as energy and mass. "In the nineteenth century, energy began to undergo a similar transformation: natural philosophers adapted a word meaning vigor or intensity. They mathematicized it, giving energy its fundamental place in the physicists' view of nature," James Gleick writes in his masterful book *The Information*. "It was the same with information. A rite of purification became necessary. And then, when it was made simple, distilled, counted in bits, information was found to be everywhere."

You could even look at the financial and organizational models that Silicon Valley perfected—freewheeling venture capital–funded companies clustering together to create a seething ecosystem of competitive, collaborative frenemies with access to risk-seeking capital. There are many different business historians who have gone about the task of explaining Silicon Valley in precisely this way. The touchstone in the genre is AnnaLee Saxenian's *Regional Advantage: Culture and Competition in Silicon Valley and Route 128*.

All these different accounts help to explain the phenomenon of a world overrun with technologies in a way that would have been unimaginable fifty years ago. But they primarily explain the technological and organizational development of the companies. There are also ideological dimensions to computing that have made their rise so powerful and unsettling. For a time, it seemed that computers and the networks they could create would lead to new types of freedom and human flourishing. They would allow new configurations of people, who could become the revolutionary intercommunal network that Huey Newton dreamed of.

People tried to make this version of "the net" happen! Some succeeded for a time, but computing did not end up freeing the people. Instead, just a few companies came to dominate the world, making vast sums of money, creating surveillance data for the U.S. government, increasing inequality, and tearing cities apart. Worse, their financial success and scale has made them largely insensitive to these consequences, and this has been the case for decades.

This was true right at home in Silicon Valley, where they created one of the nation's most significant environmental catastrophes. It's been documented extensively by Lenny Siegel, who was a major part of the anti-war protests at Stanford. Charming, sweet, and goateed for decades, Siegel has maintained an archive of dirt on Silicon Valley since he set up an activist group called the Pacific Studies Center in East Palo Alto in the early 1970s. The Center wanted to connect up the difficulties of technology workers across the world.

"When the Vietnam War ended, we did an assessment of where we were at. We'd been focusing on the war, and written stuff about other Asian countries," Siegel said. But looking around at the booming electronics industry, it dawned on them: "Oh! Silicon Valley!" The files began to pile up. "We began looking at the defense impact, international impact, and the toxics," Siegel said. Decades later, Siegel has a glorified garage behind a coffee shop in Mountain View, where he's kept thousands of documents and rigorously categorized newspaper clippings.

Silicon Valley was *the* American economic success story during the economic dark times of the 1970s, but Siegel tried to point out the problems that the industry was generating even as it was creating computing. In 1977, the Pacific Studies Center put on a conference called "Silicon Valley: Paradise or Paradox?" Three hundred people attended, and a pamphlet on the industry's costs was produced. It's a remarkable document, foreshadowing so many of the issues that continue to plague the technology industry. There was a growing jobs-housing imbalance in the Bay Area, which forced people into long commutes and terrible living circumstances. There was racial and gender discrimination baked into the industry. The tech companies were fiercely anti-union, even as they tried to automate work or ship it to Asian countries. And an industry that had pitched itself as an alternative to the old polluting industries turned out to have just buried its dirtiest secrets.

Silicon Valley processes used different kinds of powerful solvents. Chip companies held these chemicals in large underground storage tanks, in part for aesthetic reasons, so they'd look less like real factories. And as with other buried things over time, the tanks were forgotten.

In most Silicon Valley stories I've covered, the technology industry likes to separate out the triumphant tale of innovation from the unfortunate environmental and labor codas. So Moore's Law becomes one arc, and the toxic chemicals used to execute it become a separate issue. But the pollution happened right alongside the glory days. As semiconductor companies hoovered up money from the military and its contractors, taking from the public coffers, they were adding debts to the environmental commons.

In the early 1980s, the Silicon Valley Toxics Coalition published a map of the number of underground chemical spills. It showed dozens and dozens of different locations polluted by high-tech firms. This was an industry-wide problem.

The environmental remediation that has had to occur has been massive. Nineteen locations were declared federal Superfund sites both because of the severity of the problems and the intense organization of the environmental coalition cooking in the South Bay. Santa Clara County had more Superfund sites than any other county in the country. They were, perhaps as you might expect, disproportionately placed near poor neighborhoods with larger numbers of immigrants and people of color.

At the Fairchild site where the leaks were first widely publicized, the company removed the tank, then scraped the top 52 feet of topsoil from the site, removing almost 92,000 cubic feet of soil. It flushed vast amounts of groundwater through the site. At the same time, the company built a huge clay wall extending down 140 feet to contain the spread of the chemicals. All told, 147,000 pounds of toxic volatile organic compounds were removed. A developer bought the site in 1990 and redeveloped it into Bernal Plaza, home to a McDonald's and a Starbucks.

Silicon Valley, I always tell people, is a postindustrial landscape, whether or not it looks like one. There are always costs to being part of the Pacific Circuit.

14

Trauma

In that long stretch of years between the early 1970s and when Margaret became Ms. Margaret Gordon, there was little but hard work in her life.

There was the *work* work, cleaning houses or cooking food.

There was the work of keeping her head on straight as she struggled with her mental health. Just fighting off the depression was a job. The trauma seemed to be compounding. Her mom's alcoholism had pickled her. The brightness, the possibility, the movement of her childhood was gone.

There was the parenting work. Her relationship with Gordon ended, finally. She met another man, who became the father of her third son, Zuri Maunder, born in February 1977.

There was the work of surviving an Oakland in decline. The white population had plummeted from 330,000 in 1950 down to 131,000 by 1980. The Black population had gone in the opposite direction. White flight took personal wealth and commercial capital out of Oakland, and it also changed the politics of taxation. As with other cities, once white people had moved to the suburbs, they became markedly more anti-tax. In 1977, Oakland elected its first Black mayor, Lionel Wilson. He'd run the city's independent anti-poverty board for six years and served as a judge. Wilson was one of the few real links between the radical '60s—the Panthers supported his candidacy—and the more moderate business community.

Shortly after he took office, California voters passed Proposition 13, which cut the property taxes that are the lifeblood of local government.

The Oakland activist Paul Cobb recalled, "Proposition 13 was white folks' message to us that we're gonna have to do it ourselves."

As much as the face of Prop. 13 had been homeowners, two thirds of the tax savings have gone to businesses. In 1967, corporations paid 46 percent of the taxes in the state. By 1979, that number had fallen to 28 percent. Individuals, meanwhile, had to pick up the tab, increasing their share from 54 to 72 percent. "Not only did Proposition 13 help to accelerate a shift in the distribution of public resources away from older, poorer cities like Oakland," the historian Robert Self concluded, "it subsidized and masked a larger social redistribution of the tax burden, from corporate and business capital to people." The reduced tax revenue stifled the ability of cities to serve their residents and drove them toward building new commercial properties in search of the revenue to sustain services.

Meanwhile, the jobs went elsewhere. The white-collar ones went to big suburban office parks east of Oakland in places like Bishop Ranch or down near San Jose. Manufacturing jobs went to Silicon Valley's immigrants, Asian factories, or Mexican maquiladoras.

Neighborhoods were distressed, houses abandoned. Starting in the 1980s, drugs and violence began to rip apart the Black community. Huey Newton, the very symbol of Oakland's uniquely radical Black culture, had become drug-addled. Word was that Newton had been robbing the dope boys of money and drugs, and eventually, his revolutionary reputation was not enough to protect him. In 1989, he was murdered in the streets by a member of the Guerilla Family gang.

In *that* city, at *that* time, when Zuri acted out, Margaret didn't have the parenting skills to keep her cool, and she worried if he went down the wrong path, he might meet the same fate as Huey and so many other Black men. She had to keep him in line. "Whipping his ass was the only thing I knew to do," she remembered. And so she did.

Tony came back to California after he graduated from high school in 1983, followed by Jarrold the next year. They were living with Margaret. Jarrold was all grown up. Ripped, he had become a part-time male model.

Nobody could get Zuri under control, though, and as he moved toward his teenage years, their conflicts turned into a crisis. "I got into a situation where I was so frustrated with my son that I hit him," she told me one afternoon in her office. "And he went to school. I had hit him so hard that he got a blood clot in his eye and the school reported it to social services and they removed him from the house."

I didn't discover this moment in the course of deep investigation. Margaret just flat-out told me. Margaret wanted me to know. I think she needed me to know? "Sometimes it's so dark you gotta stick pins into it to see the light. You have to do the work to reinvent yourself, to restructure yourself in such a way that you know there are positives, even with the trauma," she said. "How do you use it to move yourself to a place where you come to peace with yourself? I had to grow."

In her effort to get her son back, she went to therapy with a Black Panther–turned-psychologist named Derethia DuVal month after month. She attended dozens of court dates, meetings with social workers, and visits with Zuri. "At the end, the judge said: 'Ms. Gordon, I have never had a parent like you,'" Margaret recalled with pride. "'You have not missed a therapy session. You ain't never missed a visiting day with your son. You have not missed nothing.'" After more than two years, Zuri moved back in with her, and they've had a solid relationship ever since.

Her life had been filled with trauma—and not only at the hands of the men who abused her. It went all the way back through America and straight down into the ground of Oakland. The traumas of family and race mixed with the degradations of space. "The trauma was the depression, the alcohol abuse that my mother had went through," she said. "How do I triumph myself above that—and still be engaged in all these social justice issues and caring about other folks around me? How do I not only deal with my deep internal issues but be able to take my lessons learned and give others pathways?"

That afternoon in her office, she handed me binders filled with photographs and documents. We laughed about her youth lewks, gawked at her '80s tank tops, appreciated her growth into a woman in full. There were dozens of old Christmas cards, booklets for memorial services, and forgotten flyers with inscrutable notes penciled on their backs. Here was one version of Ms. Margaret's life. Not the meticulous archive of someone who knew she would be of historical interest from birth. Not the messy profusion of a pack rat. But a thin and somewhat haphazard chunk of what it was like to survive pending revolution.

As I went through it, piece by piece, I found a set of papers doubly wrapped in plastic ziplock bags. They were court documents laying out the administrative history of Margaret's case with social services. The papers confirmed what Margaret had said about her record in the reunification process. She had enthusiastically cooperated and supported the process "in every way."

But they also detailed the incident that led to Zuri being pulled out of Margaret's care.

> The minor has suffered, or there is a substantial risk that the minor will suffer, serious physical harm inflicted non-accidentally upon the minor by the minor's parent or guardian, in that:
>
> 1. On or about October 11, 1989, the minor was hit by his mother, Margaret Gordon, with a shower hose on the back, left cheek, and eyes, resulting in a red and swollen eye;
>
> 2. The minor has been previously hit by his mother with a belt and a stick;
>
> 3. On or about February 1989, the minor reports his brothers held his arms and legs while his mother hit him with a belt and then with a stick;
>
> A. The minor reports that he was bruised as a result of this beating;
>
> 4. On or about October 11, 1989, the mother ordered the minor to do 200 push-ups when he returned home late and she also denied his supper.

Margaret had told me what happened, but as I looked at the words there in black and white, I felt a sickness descend on me. I could not help but see Margaret standing over her son, a child, twelve years old, raining blows down on him. Both my parents were abused by their parents. The thought of my own kids getting hurt is physically unbearable to me.

But how could I judge? I knew nothing of what it was to send a Black boy into the Oakland streets at the peak of the crack crisis and the long twentieth-century crime wave. Violence was everywhere.

I knew that the story had a happy ending. I knew she had put in seven years working at the West Oakland Health Center in the mental health department, learning and relearning the lessons she needed to make herself whole. In her own words, she gained "maturity and self-reflection and the ability to forgive myself for those mistakes."

I knew that she would be reunited with her son, and that he would be, decades in the future, going to her seventieth birthday party with his own son, that this grandchild would be gently laid in a chair, and draped with a jacket to sleep peacefully. Mother and son would rebuild their relationship on solid ground, away from the toxicity and abuse that she'd taken into her body over the decades and then meted out on her child. I knew that this was the crucial moment in her development as a human being, and that she was a woman who had been subjected to the riptides of her race and class and gender and generation, forces that destroyed so many others.

Her survival alone was a triumph. Her ability to transmute all that pain into a new, reconstructed self, who would relentlessly work to heal her community was borderline miraculous.

I knew that I was not in a position to pass judgment, coming from where I was coming from and with all the privilege that I have had doled out to me.

I knew all that. And still.

I sat on the couch in my office with the binders on my lap. The daylight leaked out of the room, light outside bending orange. The documents pushed me lower and lower. I couldn't think, the sadness was overwhelming. It felt like something was being stuffed into my body. This trauma, even at two steps' remove, still had so much dark power pulsating within it. The feeling was thick and suffocating and oily.

The abuse, the trauma, the collapse of Black Oakland pushed and pulled by American racism and technocapitalism, all the brilliant people out there scrubbing toilets to survive, poisoned by the paint in their walls and the air they had to breathe to live. And through everything, they were told that it was their fault, that it was their culture, the same culture that has forced America to deliver on its promises of freedom, that has produced so much of American music, that lies curled inside all our tongues, that is the American export that communities around the world embrace and love as a distinct contribution to human civilization.

All of that kept pouring into my body as darkness kept falling outside. When I was able to move again, sloppily, as if I did not have full control of my limbs, I switched on a light, put the documents in a filing cabinet, and couldn't work on this book for a month.

The ultimate privilege for all non-Black Americans is not to take on this suffering as our own, to not see *ourselves* enslaved, *our* families ripped apart, *our* life under apartheid, *our* babies shot by police, *our* communities bulldozed, *our* aspirations thwarted, *our* traumas passed down from child to child to child.

If we did, what, then, would our responsibilities become?

PART III

Birdsong

Phil Tagami

15

The Shadow

All that you touch
You Change.
All that you Change
Changes you.
The only lasting truth
Is Change.

—OCTAVIA E. BUTLER, *PARABLE OF THE SOWER*

If there is a before and after for West Oakland and for Margaret Gordon, it is October 17, 1989. As Margaret convulsed with grief over her son becoming a ward of the state, a 6.9-magnitude earthquake hit right at rush hour, 5:04 p.m. It was a catastrophe. Thousands were injured, and dozens died, mostly in West Oakland, where the freeway that had fulfilled the long-held dreams of white segregationists—the Cypress Structure—collapsed on itself, pancaking cars between upper and lower decks.

There were stories, of course, about Black residents looting the bodies of the people trapped on the highway, because that's how a lot of people saw West Oakland, but the opposite was true. People in the neighborhood sprang into action and saved lives, as described in a stunning episode of the podcast *East Bay Yesterday*.

Raven Roberts, a local resident, lived right next to the freeway. He remembered the terrifying moments just after the quake. "There was a cloud of dust, I would say, maybe forty feet high. You couldn't see the freeway structure anymore. All you could see was the cloud of dust moving towards us. And then out of this cloud of dust, this man runs out, his eyes were as big as saucers, and he yells: the freeway is falling, the freeway is falling. I looked at my wife and I said, 'Come on,' and we ran into the cloud of dust."

The couple arrived at the scene as first responders. Other damage would become known shortly. A fire in the Marina. A section of the Bay Bridge collapsed. The Embarcadero Freeway destabilized.

"When you first got into the cloud, there was no sound. It was like standing in a brown, moving fog bank. Thick," Roberts said. "Then my mind starts to hear horns blowing, and the horns are blowing louder and louder and louder. And my wife says, 'There's a man there.'"

They began to help free people, doing their best to save lives, working alongside the fire department. It was a horrible and heroic moment in their lives.

Afterward, light shone in places where it had not for decades. Noni Session was a freshman at McClymonds High School at the time. Her father, Major Session, Jr., ran a grocery store in the shadow of the Cypress. It was just shouting distance from her grandmother's house on Ninth Street and a couple blocks from her Aunt Glory's spot on Center. Noni had grown up bouncing around that triangle for as long as she could remember. It was a family business and everyone pitched in.

The Cypress Structure was a constant presence in her life.

"The memory is very vivid. I can feel the constant ground vibration and I can hear the constant roar, even right now," she told me. She remembered being eight years old and standing there looking up. "It's the middle of the day and it is dark, really gray. I just thought that's what the weather was," she said. "It was always kinda cold and loud. Whooosh! Whoosh! The cars right up above your head." In a photo taken by California Department of Transportation (Caltrans), you can just barely make out her father's store, a small white building, dwarfed by this "giant stone angry looming structure," as Noni put it.

Then came the earthquake. It was a while before her parents let her return to see what had happened to this place where she had spent more than half the hours of her youth. By then, Caltrans had mostly disassembled the freeway and it was just sitting in piles in the middle of the street.

She set foot outside her grandma's house, about five houses down from the Cypress. "I stepped forward and it was like a pause. I don't even know how long the pause was. It was eerily quiet. It was super sunny, BRIGHT. I can see the light straining my eyes. I was in awe," she said. "And I stood there and this bird starts tweeting. That's a BIRD! It was like I had never been in that place before."

The very highway that was a monument to destroying West Oakland had come down, deus ex machina.

And the question was: What would be rebuilt?

With Margaret's permission, I got in touch with Derethia DuVal, the woman who had led Margaret out of her depression and through the reunification process with Zuri. DuVal agreed to talk with me. She lives in West Berkeley, just a few blocks from my old office. A string of pots stuffed with succulent cuttings ran from the yard haphazardly up the stairs onto the porch. Classic East Bay, unbothered.

I knocked and an older woman with locs answered the door. She was friendly, almost maternal, unpretentious. Books covered literally every surface of the bright home. She was not someone who ran around the house tidying up before a journalist arrived.

We cleared a spot for my computer on her kitchen table and talked about a recent event in Hunters Point, tying together the major uprising there in the late '60s with the gentrification of today. Then we got down to Margaret. I handed her the documents, which she looked over as I explained my understanding of what had happened. Did she remember Margaret's situation? "I remember the feeling," she said. She looked back at the papers, evaluating. Then she stopped and dropped them on the table.

"She was being fucked by the system," she said, and burst into laughter.

Her tone changed. I could imagine her teaching at San Francisco

State, as she had. "We've talked about housing and displacement, but racism has a lot of forms. The mental health and juvenile justice and criminal justice systems have a bias, and the bias is white supremacy," she said. "The overlay of everything we exist in, that we experience, that we feel, whether we know it cognitively or not, is the doctrine of white supremacy. And the part that's not spoken about is that the doctrine of white supremacy says that African Americans are inherently inferior people. If it stopped there, that'd be one thing, but juxtaposed against that is that white people—gentiles—are inherently superior. That is the unconscious, as Jung calls it, the shadow, that is part of all of our socialization, those who have come up through the continental United States. This is ours. This is what we share. I'm saying that all to say that the difference between her seeing me and her seeing somebody else is that I saw Margaret's frustration. I saw Margaret, first of all, as a Black woman who was doing the best she can with what she had. Period. All the history, I could see that in her. And Zuri having some issues that were diagnosed, not diagnosed, misdiagnosed, all of the above, and basically, there were no resources for people who were poor. There weren't in 1989 and there aren't now. And most of the counselors and therapists and probation officers came from a world that was nothing like the people they were monitoring. And so it's not that they were bad or uncaring people. It just is, it's in the shadow. So, when you come to the system, you automatically come with two strikes."

She continued. "Margaret was always a strong person, her personality. And then we have our stuff, too. But I saw her. As a person. And I knew she would be the best person to raise her child. The institution doesn't have any love for Black children. I did believe in her. I knew she could do it."

DuVal's perspective was honed over decades, first as a child of Harlem, then as an organizer with the Congress of Racial Equality and the Student Nonviolent Coordinating Committee, and finally as a member of the New York chapter of the Black Panther Party in the late 1960s. That's when she began to become an African psychologist, as she says, through the reading and political education that the Panthers required. "I'm a baby boomer. We were the transition movement, we were a paradigm shift," she recalled. "And it was all of us—Black, Latino, Asian, Fijian, people all over the world were coming out of the yoke of colonialism, realizing their identity."

After coming to California, she fell in with the remnants of the Black Panther Party in Marin City, who would later become famous for taking care of Tupac Shakur. They wrote grants to set up the first mental health crisis line in Marin City. The line would ring through to a different former Panther each evening. Eventually, the operation turned into the Marin City Multi-Service Center.

Then she got pregnant and realized she needed a real job. She started at the College of Marin, then bounced to Sonoma State, and SF State for a master's. "I was always pan-Africanist," she told me. "That's what frames my life." As she was coming up, the Association of Black Psychologists was beginning to question the field's premises. "If you're economically enfranchised and privileged, your inner life is the most important thing about you," DuVal said. If you're not, then what is happening outside the boundaries of your mind can be as important to your mental health as any constellation of emotions.

After graduating, she went to work at the West Oakland Mental Health Center under Dr. Isaac Slaughter. "Isaac Slaughter, he was a race man," DuVal said. "You look at the world through Black lenses. We just were race people."

She shook the papers from social services in the air.

"Looking at these papers and looking at Margaret, it's like: Yeah, that's what they had to say about her. And I already knew what they had to say was bullshit. I just had to see how serious is Margaret. She had hit rock bottom and there was nowhere else to go but up. It was in her. Institutional racism beats you down. It beats you down from the beginning when you have ignorance, you have poverty, you have a constant mismessaging about who you are. And generations of that," she continued. "To me, we have to look at it, when people inherit wealth, they also inherit poverty. We all inherit our history."

Of course Margaret threw herself into organizing the community, that was of a piece with African culture. "I think service is part of our healing and is part of who we are. We're a fish out of water in this culture," DuVal said. "This culture believes in retribution and revenge, and we're a redemption culture."

In DuVal's eyes, what Margaret learned through the hard work of getting Zuri back is what gave her insight and courage. "Self-knowledge is the highest form of knowledge. That is the core of African psychology. When you know yourself, you know other people. She knows Margaret. She knows why she's doing what she's doing," DuVal said. "She doesn't need to be loved by other people. They need to be loved by her."

In Oakland, within days of the earthquake, Paul Cobb, Bill Love, and Ralph Williams had formed the Citizens Emergency Relief Team. Their goal: to force Caltrans not to rebuild the freeway along Cypress Street. In the 1950s, Cobb's father *and grandfather* had hired the future Oakland mayor Lionel Wilson to battle Caltrans over that same damn freeway. "The argument was that it would divide the community and the bad air would give respiratory diseases to ordinary schoolkids," he said in a speech at an event for Margaret. "And that prophecy came to be true." They were not gonna lose again, and they got eighteen cities in Alameda County to pass resolutions against rebuilding the freeway.

Meanwhile, a new group of environmentalists led by Chappell Hayes argued that no freeway at all should be rebuilt in West Oakland. "Freeways have caused high cancer rates in the communities alongside them and high incidence of lead in the brains of our children living in these communities," he said. Hayes was part of a growing movement of what came to be called environmental justice. "When you say environment, it evokes images of the spotted owl and the ancient forest," he said in a 1993 interview. "I'm certainly concerned about those issues, but it doesn't usually connote the urban environmental issues that I'm usually involved in." Hayes was married to Nancy Nadel, a future city council-woman, and they made a powerful interracial team.

During those years, Hayes met a young man named Phil Tagami, who began to involve himself in civic efforts. Tagami volunteered on Hayes's unsuccessful run for city council. Hayes mentored Tagami on city affairs, teaching him how to work with data to see patterns in the economic and political development of Oakland. "He was very real. He had no problem calling bullshit or breaking his foot off in your ass," Tagami said. "But he was very sensitive and thoughtful." In one photo, Tagami and Hayes pose together with a group of people who'd done a street cleanup, a multigenerational, multiracial landlord and tenant community improvement group.

The combined efforts of the different West Oakland groups forced Caltrans to rebuild the freeway on the very edge of the neighborhood. They won, that is to say, the reunification of West Oakland. "If they had tried to rebuild it we would have all been out there and they would have had to mow us down," said Ellen Wyrick-Parkinson, a longtime community leader.

Hayes, though, would not live to see his environmental vision come to fruition. His life was cut short by an aggressive form of cancer. "There were moments when he was diagnosed with pancreatic cancer where not a lot of people knew what was going on and he was wrestling with the fact that he wouldn't be there to support his family," Tagami remembered. "And he never complained. He was dedicated to what he was doing until the end."

Margaret arrived in West Oakland just as Hayes was passing on. She joined up with the Coalition for West Oakland Revitalization, which came together to lay out a new vision for the neighborhood, in a post-earthquake, post-Cypress world.

"October 1989 the earth shook with a furious rage and forever changed the destiny of West Oakland," the group's leader Queen Thurston wrote. "In the space of a few agonizing seconds the Cypress Freeway, a literal and figurative symbol of a divided community, came crashing down. Though there was great turmoil and pain, beneath the concrete, this tragedy also brought with it a seed of hope and beauty: West Oakland could be made whole again."

CWOR's first report laid out the challenges West Oakland faced in 1994. Official unemployment was over 20 percent. Median household income was 60 percent of the Oakland average, which was itself a fraction of that of wealthier Bay Area cities. Ninety percent of West Oakland schoolchildren were in families receiving Aid to Families with Dependent Children. Teen pregnancy was as high as the lead exposure. A million feet of industrial and commercial space lay vacant. Factories and homes sat side by side. The housing was old and dilapidated. The air was bad, the soil contaminated. Streets were noisy, dirty, and dangerous. Diesel trucks drove all over the neighborhood.

Yet their statistical portrait did not *only* describe a decimated community but a financial opportunity for real estate speculation. This was a place, after all, that was mere minutes from both downtown San Francisco and Oakland. Chris Roberts, Raven's wife, described a massive transformation in how people saw West Oakland after October 1989. "People realized that we weren't horrible people," Roberts said, "and they realized the freeway is down now, so this is great, great property."

To outsiders, the housing stock had a pleasant New Orleans vibe. Most important, in 1990, 70 percent of West Oakland was Black, 10 percent was white. As nearly always happens in real estate markets, if more white people moved into the neighborhood, prices would rise. "The area is attracting a new population of individuals seeking a closer proximity to Oakland and San Francisco, affordable prices, and architecturally attractive structures," CWOR noted. "These factors can potentially lead to 'gentrification,' defined as what happens when one income (or cultural) group replaces another group in a neighborhood." Given its excellent location, if the neighborhood got better along any dimension—crime, environment, transportation, anything—it was almost certain that more white, higher-income people would move in. As in so many places, it would prove difficult to "improve" the neighborhood without displacing its people.

By day, Margaret cleaned houses to support her family. It was not glamorous work, but it helped her make a fortuitous connection. One of her clients was Michael Herz, an OG environmentalist who founded the organization Baykeeper to monitor pollution in the San Francisco Bay. When it caught someone, it took them to court, becoming a kind of third-party enforcer of environmental law.

When she took breaks from cleaning Herz's house, she began to leaf through the environmental magazines he subscribed to. During the early 1990s, they were filled with an emerging civil rights movement that would transform the way people thought about "environmentalism."

Out at the port, the ILWU and the employers had their choreographed battles about labor's share of profits. The traditional environmentalists, too, had been fighting battles to protect the natural parts of the bay. In particular, they tied up dredging for years and years with legal wrangling. There were tables—physical and metaphorical—at which these various groups sat to work out deals.

What about the rest of the people? The people living next to the port, breathing in the diesel fumes from the trucks, inhaling the exhaust from the ships burning bunker fuel? They had to take on some of the costs of the business that put money in ILWU members' pockets and into the bank accounts of the companies they worked for. But where was the people's cut for taking on the environmental burdens?

The Port of Oakland, through some strange historical circumstances, was owned by the city, but it wasn't run directly by the mayor and city council but by a separate institution, the Board of Commissioners. They were appointed by the mayor and voted in by the city council, but they represented the business interests of the port, and always had. Anybody from the community was on the outside of the process and would have to make their presence felt some other kind of way.

Margaret learned the history of what became the environmental justice movement. Locally, of course, there was the example of Chappell Hayes. Nationally, the histories usually began in the early 1980s. Warren County, North Carolina, residents teamed up with the United Church of Christ Commission for Racial Justice to fight a toxic waste dump from being sited in their community. They lost their battle, but a new way of talking about and organizing for environmental protection sprang forth.

Placing polluting facilities in Black communities had largely just been assumed by planners. So, of course, planners would site the most hazardous facilities in the places where land was cheaper, buildings were old, and political opposition less powerful. Time and again, in the Bay and around the country, that meant toxic facilities ended up in exactly the places where Black people had been forced to live. As we saw in Part I of this book, I. S. Shattuck suggested siting industrial facilities in what was an incipient Black neighborhood to crowd out the people living there, leaving it ready for industrial redevelopment.

At night, Margaret went to community meetings, led subcommittees, and sunk more deeply into community leadership. While the West Oakland community won their fight to reroute the Cypress, the process of building something new in its place proved harder. CWOR pushed for and failed to get an "African marketplace" established in the place of the Cypress structure to support local Black businesses. In the end, the neighborhood would not have a new parkway for fifteen years, at which point many of the residents who started the fight had already left or been displaced, and the finished path wasn't what they wanted out there anyway.

The environmental justice movement, particularly in the early days, wasn't just a crucial observation about the relationship between power and pollution. It was *also* a corrective for environmentalism, which had come to mean protecting "nature." Rachel Carson, who wrote *Silent Spring*, provided the intellectual heft and direction for many activists. Ecosystems should be preserved. Land should be kept from development. Toxic chemical use should be decreased to save our flora and fauna.

This was an important intellectual achievement. It required all kinds of people to consider a new proposition about the earth: it wasn't only humans who mattered. We were part of larger systems—biological, climatic, hydrological. The earth was not just one big ball of raw materials, of "natural resources"; it was a living entity built of communities of organisms living together in complex ways, humans included. Our behavior was steamrolling other creatures and, in many cases, destroying the natural systems on which our own lives relied, dirtying water, polluting air, and all the rest. And of course, there was the experience of "being in nature," which is close enough to a religious feeling that it creates powerful devotion to the cause across demographics and sometimes even political orientations.

As it developed historically in the United States, environmentalism tended to focus on the concerns of white, wealthier people. The new Environmental Protection Agency issued a report on the "urban environment" as early as 1972, but the mainline environmental organizations such as the Sierra Club focused on animals and on what passed for wilderness. Efforts succeeded at preserving open space in Marin County, in the hills east of Oakland, and around the small cities of Silicon Valley.

Of course, I feel some gratitude to these people, who did, in fact, preserve the bay as a functional ecosystem and beautiful component of the world. Remembering the creation of the San Francisco Bay Conservation and Development Commission, I find it astounding that the captains of industry, who saw filling in the bay as something like manifest density, were turned back by a woman-led emergent movement. In less than a decade, those environmentalists had wired in the power of the state to protect the commons! The Bay Area we know is literally unthinkable without that contribution.

But it had an ugly side, too. For example, Silicon Valley communities, citing environmental concerns, made maximal use of their local governments to enact highly restrictive planning policies that made building new housing very difficult, without comparably restricting the dynamism of its powerful businesses. Environmental regulations placed huge chunks of the region off-limits to development. By enforcing a strict, low-density suburbanism within a booming manufacturing region—prices shot up. "The end result, largely, was prioritizing conservation and open space over socioeconomic diversity," wrote the environmental historian Jason A. Heppler. "Particular kinds of open spaces were prized over others—an almost quintessential NIMBY ('not in my back yard') politics played out across the Valley landscape."

Environmental justice advocates were working in places that had no open space to preserve, that were in desperate need of affordable housing, that would never feel naturey.

Margaret had direct experience with the overburdened community. She had asthma and so did Zuri. When she went into his schools as a parent advocate, every teacher had a box of inhalers for the kids in the classroom. It wasn't hard for her, as someone who lived in the community, to imagine why this might be. The community had been saying for literally fifty years that the pollution coming from cars, ships, and most especially port trucks was sending particulate matter into the lungs of children, opening them up to respiratory diseases. The containers did not move themselves, and the trucks that powered them to and from the warehouses were a crucial drag on the health of the community.

Hers was an environmentalism wrenched from the eternal realms of beauty and nature and God's creation into the earthy history of humanity, bodies, and economic systems. "Who's gonna live?" Ms. Margaret explained once. "That's what it comes down to: Who's gonna live? Not how you gonna survive, but who's gonna live through these different phenomena and changes that's happening on Earth?"

The environmental justice movement would not let the powerful pretend that race was not a component of land use decisions, even if white lawmakers, mayors, companies, or citizens would have preferred not to see how race structured the pollution burdens of communities.

The American EJ movement immediately linked up with people from across the world. After all, if the companies of the Pacific Circuit were willing to pollute their own backyards, as they did, just imagine what they were doing in Hong Kong, Malaysia, and Korea. Multinational corporations, run by white executives in wealthy, unpolluted communities with plenty of "open space," were more than happy to sacrifice other people's homelands.

The internationalist synthesis took official form in the "Principles of Environmental Justice" agreed to at the First National People of Color Environmental Leadership Summit in 1992. This was no longer a local movement dedicated just to toxic siting. It embodied the transnational radicalism that Huey Newton had imagined with his "revolutionary intercommunal" politics. This was a continuation of the anti-imperial movement, a thread of the nascent anti-globalization movement, and even a step toward a post-capitalist economy.

Activists who had seen the '60s radical period rise and fall could vibrate at these frequencies. A delegation of Asian Americans from the Bay Area had gotten a grant from the San Francisco Foundation to go to the summit. They came back fired up and ready to get to work. "It was so dynamic and real," recalled an attendee, Pam Tau Lee. "You could just feel the energy and the power." They decided to found the Asian Pacific Environmental Network, which got to work organizing Laotian refugees in Richmond and Chinese immigrants in Oakland. They protested the fossil fuel giant Chevron up in Richmond and a small semiconductor company, AXT, for what they claimed were exploitative labor practices. They raised up a generation of young Asian EJ activists.

As the Communist bloc collapsed and neoliberalism became the order of the day, environmental justice offered a generative alternative to simple survival under increasingly precarious market conditions. The activists targeted the development philosophies and processes that had guided decisions in so many poor places from Prescott to the Philippines. The most important principle was that the people closest to the environmental problems should be the ones to speak on them. No more would somebody from downtown or America or the World Bank or Washington, D.C., determine the fate of places they'd never even seen. Residents wouldn't just find out about the plan at a public meeting and have to react. They should be the ones writing the plans themselves, even if they didn't have the capital to execute them.

They explicitly rejected the individualist market philosophy that had come to dominate American life. Through the strangeness of the 1970s, the historian Daniel T. Rodgers identified that our society entered "an age of fracture," in which "the aggregate aspects of human life" lost out to "its smaller, fluid, individual ones." Human nature itself got reconceptualized to emphasize "choice, agency, performance, and desire," as "imagined collectivities shrank," Rodgers wrote.

This was a radical ideology, but it appealed to the individualistic strain in American life, and it was deeply, spiritually useful for the corporate elites who emerged as imperial globalization cracked open new markets. What had been called morals were actually just tastes. By pursuing their own interests, people were doing *exactly the right thing*. And in any case, people *always and exclusively* pursued their own economic self-interest. The only true good was market efficiency.

In that context, the environmental justice movement was one of the few bulwarks against what we might call neoliberalism. In an age of fracture, it would opt for communal repair and healing.

16

Predatory Inclusion

The deepest, most haunting forms of American racism work through property. Before and after slavery ended, residential segregation was the tradition of the nation, even as it passed into and out of the law of the land. As a shelf full of books can attest, the U.S. government knowingly and intentionally built a racist real estate market, all while disclaiming responsibility for tinkering with the market at all.

Banks, in the old imagination, were places of rectitude and restraint. Some guy in a fancy suit—sometimes he was fat with a big belly straining against a vest, other times thin and severe, glasses perched on nose—sat in an office somewhere. Nearby, there was a vault. As any bank robber would tell you, that's where the money was. The money came from people putting their hard-earned dollars in there, standing in line to deposit paychecks, rolling quarters into neat packets, bringing in the cash from the bar. That money would then be lent out to people to do the works of the community. An olive mill, a new car, and, of course, a mortgage for a home.

In the white middle-class version of this story, it's all great. A nervous young couple heads in. The man has a dependable job down at a factory or in a downtown office building. The banker eyes them, judges them *creditworthy*. They shake; a SOLD sign gets plunked down on their ranch-style house in the suburbs; the Wonder Years ensue.

And you know, for tens of millions of white Americans through the decades, that really is a decent approximation of how it worked. The federal government shaped and stabilized the entire residential real estate market so that that exact couple would have that exact experience at their local savings and loan. It made them good Americans, anti-Communists, property owners invested in their city, their neighborhood, their block. The percentage of Americans who owned their homes increased from 44 percent in 1940 to close to 63 percent twenty years later. That's some effective policy work right there.

This effective, morally dubious system fell under pressure from civil rights activists. There were many turning points along the way, but none was more symbolically important than the Fair Housing Act of 1968, a homage to the murder of Martin Luther King, Jr., and the result of decades of pressure by Black Americans and their allies. The legislation officially ended racial discrimination in housing.

But that did not mean that Black Americans experienced full and equal inclusion in our real estate and mortgage finance system. Keeanga-Yamahtta Taylor's work on "predatory inclusion" has shown that the high liberalism of the 1970s kind of, sort of attempted to redress the problems of the previous regime of mortgage finance, then abandoned the effort when it proved difficult. Black people got increasing access to mortgage finance, but on worse terms and for worse housing. Late twentieth-century urban home ownership was not a simple upward spiral, as it had been for the suburban white middle class of the mid-century. And that is certainly an important part of what made life in West Oakland difficult.

Many histories focus on the breakdown of the spatial barriers Black families faced in finding housing. Redlining worked in concert with other practices such as racial covenants (explicit and implicit) and the "steering" of different racial groups into the neighborhoods where they were "supposed" to live. Upwardly mobile Black people often broke these barriers from the 1950s onward. Integrated neighborhoods became theoretically possible, and some even came to exist in Oakland.

But just a few months after the Fair Housing Act, the Housing and Urban Development Act of 1968 passed, creating the possibility of the mortgage-backed security. Slowly, it would create new configurations of race, risk, and place that would help open cities to global capital. Gone would be the local banker. Gone would be the humble people with their paychecks and quarters. Gone would be the block-by-block spatial risk analysis. In their place would be a global pool of money, circulating around the world seeking return, and powerful computers that bundled and sliced mortgages into complex securities that let people buy into the cash flow from American real estate.

Banks didn't want to lose the money they lent out. So, in the previous eras of the American mortgage, they controlled that risk by, as the sociologist Martha Poon put it, "screening." That is to say, people seen as too risky were simply denied credit. The bank said no. As we saw in earlier chapters, a key factor in denials was a neighborhood's racial composition. The preference for white, racially homogenous areas extended to the very marrow of the mortgage finance system. The whole thing was built around spatial analysis—à la the Chicago School. *Where* the neighborhood was told you whether you should lend there. The West Oakland resident Bruce Beasley told me that he actually *saw* a red line circling the neighborhood on a banker's map in the 1960s.

Even as banks and thrifts stopped being able to use race explicitly in their calculations, it wasn't difficult to construct proxy measures, as housing activists pointed out. The money stayed away, which led to dilapidation, lower prices, and the overall decay of credit-starved neighborhoods.

Mortgage-backed securities and the "collateralized debt obligations" that Wall Street computers built out of them had a different approach to risk. The individual mortgages in a CDO were pulled from all over the place, and given this, purchasers could not base their analysis on a map. Instead, risk was keyed to individual borrowers' credit scores and other attributes of individual loans. Furthermore, risk was not even a bad thing to these modelers, as long as they could extract enough value to beat the number of defaults, whatever that number was. The people buying CDOs often had no idea what was in them, let alone who the borrowers were or where the properties were located.

There was real hope that the end of redlining could open up more credit for "underserved" communities and people. There were even explicit targets for these things built into various pieces of legislation under the rubric of "community reinvestment." And the research shows that this did work to a limited extent.

But redlining did not only affect Black people. Redlining in both its formal and informal eras kept lots of white people away from the places where Black people lived. When those restrictions went away, white people seeking good deals on real estate began to buy in Black neighborhoods. Lenders, too, were under pressure to extend credit to formerly redlined neighborhoods. They found that a great loan in such a place was to a high-income white or Asian person in a largely Black area.

So Black people's increased access to credit was *swamped* by non-Black people's access to formerly Black neighborhoods. This combined with rising immigration and declining big-R Racism to create massive urban change. Credit flowed into the central city, most of it in the hands of people who had not lived there before.

This is the primary mechanism of gentrification and its result is obvious: the massive displacement of Black people from formerly segregated, redlined, and once overwhelmingly Black neighborhoods. It didn't happen all at once. It didn't happen the same way in every city at the same time. But this is the arc.

Neighborhoods like West Oakland had been left to decay for decades, beset by ill-fated, halfhearted government programs. Then, these same neighborhoods were opened to the rest of the market as if the market itself could effect reparation rather than finding a new shape for exploitation. Given the income and wealth differentials between ghetto residents and those outside those walls, *of course* newcomers would have higher credit scores, more money for down payments, and the ability to take on remodeling projects.

In 1990, in West Oakland, there were 18,000 Black residents of the neighborhood. And yet, just 13 (13!) Black people received a total of $1.3 million in loans to buy homes there. Even though there were fewer than 1,800 white residents, they got 20 purchase loans worth $2.7 million. Asian Americans, also less than 10 percent of the neighborhood population, got 24 loans worth $2.3 million. These changes would accelerate through the '90s, helping to catalyze racial and class change. The Black share of the population fell from 77 to 65 percent from 1990 to 2000.

Just to spell it out: if you got rid of the red line, but not the harms inflicted by institutional racism, the nearly inevitable result in Black neighborhoods was what came to be called gentrification.

And then came the subprime mortgage boom.

As the Asian countries' export economies grew, they began to "recycle" their profits back into the American economy through buying dollar-denominated assets, regular bonds as well as other instruments like mortgage-backed securities. As the dynamics spiraled upward, foreign capital generated a huge pool of credit for American homebuyers. That led to a long boom in housing prices, which gave people with stagnant wages a way to continue buying the products that kept coming over from Asia—Japan, Korea, Taiwan, China. It was perhaps not a virtuous circle, but it was an important loop in the Pacific Circuit.

The U.S. economy, which had been stuck in a rut for years, began to grow faster than those of other developed countries, strengthening its relative position. The era of deregulation allowed American corporations free rein in the United States and increasingly around the world.

Finance, insurance, and real estate became wildly profitable. Even the great manufacturing companies opened up financial wings and began to rely on them to inflate their profits. While mid-century elites depended on local workers for their wealth, the new financialized elites largely did not. The institution of the big corporation—which imperfectly united the interests of workers, managers, and financiers—gave way to an unbridled capitalism represented by the private equity raiders. The leveraged buyout used the new capacities of finance to take formerly productive companies and sell them for parts, jettisoning the financial arrangements, such as pensions, that had undergirded the mid-century corporation.

And, of course, there were those mortgage-backed securities, on which so much came to rest. They got more and more complex as the computing power necessary to create and price them increased. The data necessary to evaluate people for mortgages was expanding and being crunched by the same new vast computing power.

If you weren't looking too closely, the news seemed extraordinary. For almost all of the '90s, the total amount of mortgage credit flowing into West Oakland—for all kinds of mortgages—never exceeded $30 million, but it was rising. Most years, it was between $10 and $15 million. The average loan size was fairly stable at around $100,000. And by the long-established logic of real estate appraisal, the more white people moved in, the higher prices were likely to move. That *could* be good for Black homeowners, many figured. By 2000, Black homeownership rates nationally topped 47 percent, a substantial improvement from the 1960s.

Then, quite suddenly, *there was too much money.* Lenders made $23 million of refinance loans to Black people in West Oakland in just 1999—more than all the money they'd lent for home purchases in the rest of the decade. The terms of these loans, following the predatory inclusion model, were usually not good. Often people who owed little on their homes—the most stable residents in the neighborhood—were offered big "cash out" loans that left them on the hook for higher payments down the line. Sometimes there were balloon payments or other clauses that had been squeezed out of the standard American mortgage back in the 1930s.

For regular people, this was often financially ruinous. Noni Session's mother lost the house that she'd grown up in exactly like that: a bad refi with a balloon payment. It was happening all over the place. There were 499 foreclosures in West Oakland between 1995 and 1998, almost double the total number of loans to Black residents during the same time period.

Worse, many borrowers in these areas could have qualified for better loans, but banks targeted the very areas they'd long ignored with the worst products. Scholars coined the term "reverse redlining" to describe the way that loans were flowing into previously verboten communities.

Noni was lucky. Her uncle was able to buy the house and keep it in the family. But many of the former residents of places like West Oakland ended up in industrial suburbs like Antioch or Pittsburg or even farther out in Stockton and Sacramento. The Bay Area would become a leader in what the urban researcher Alex Schafran calls "resegregation."

17

"Free" Land and "Free" Money

In 1995, the Department of Defense slated the Oakland Army Base for closure. Ron Dellums, former War on Poverty official in Hunters Point and future Oakland mayor, was a powerful congressman with socialist leanings. He'd been arguing for a reduction in military spending for years. The military's Base Realignment and Closure commission paid him back by moving to shut down every facility in his district, including the Oakland Army Base. That touched off a process that has occasionally inspired and often infuriated the community.

The basics of closing down the Oakland Army Base were simple: the military would turn over its 425 acres to the city sometime after the base shuttered in 1999. But what should Oakland do with it? The land became a canvas for the community's dreams in the ferment of those post-earthquake years. The military, Ron Dellums, the city of Oakland, and everyone else promised the community that the base closing might be tough, but it presented an opportunity. Here was prime real estate at a time of new capital flowing in! Margaret, of course, joined the group of citizens who made up the West Oakland Community Advisory Group to the base closure authorities.

It was destined to be messy. This was free land on the shoreline of San Francisco Bay. There were buildings out on the base. Huge warehouses, a nice-looking administrative center, a bowling alley. The land was directly adjacent to the Port of Oakland, with substantial harbor space. The Navy had also closed a supply base nearby, so the Port of Oakland, which had been struggling to compete with the ports of Southern California and the Puget Sound, suddenly could imagine new growth.

Not only could it get new terminals but it could put together an entire logistics hub on the acreage.

And yet the site was, for most intents and purposes, a wasteland. All the evils of twentieth-century chemistry, from fossil fuels to toxic chemicals, had been spilled all over the base for decades. Everything would have to be remediated, which would require huge sums of money.

WOCAG tried to balance the desires of different community groups with the direction of the city economic development team, the port, and the Oakland Base Reuse Authority. They thought they had real power, and wrote a preliminary report in 1998 that included land for the port's maritime use and a workforce development campus with training and placement for West Oakland.

Federal legislation required that homeless services be given top priority for making use of closed military facilities. It seemed that Oakland's Homeless Collaborative, a coalition of community groups that came together around the base effort, would get eight buildings, totaling more than 229,000 square feet for workforce development, a food bank, a child-care facility, and 52 units of transitional housing. There'd be local hiring provisions and a fund dedicated to putting money to work in West Oakland. It was a triumph of community planning that took several years and the dedication of dozens of people from West Oakland. WOCAG got written up as a model of inclusive community planning.

Then WOCAG's carefully constructed plan got blown up. The San Francisco Bay Conservation and Development Commission, which was and is the reigning authority on the waterfront, stopped the plan in its tracks. It said that it had already determined that the entire former Army base had to be used for maritime purposes. It was quite a turnaround for an agency that been formed from the environmental movement to save the bay.

The decision was a disaster for local community groups. "At the end of the day," Margaret noted, "the Homeless Collaborative got X amount of dollars, eleven million dollars or something, and they told them to go away. They got no property." The Collaborative floundered from that point onward, ending up deep in debt to the federal government.

The base itself went into a new phase of planning. By 2003, the work had been thrown into the hands of the city's redevelopment agency, which could borrow against the future tax revenue of the area. WOCAG was revealed to be yet another advisory board with no power. People had spent seven years of their lives working on the base. Nothing really came of it, except the relationships they built with one another.

The city would do whatever it wanted with the land. Ultimately, the city and the port each got roughly half the acreage. The port began to make plans for its portion, while the city more or less cleared the way for almost anything to go into its piece.

←···   ···→

In West Oakland, the Hewlett Foundation, created from Bill Hewlett's chunk of the HP fortune, decided to invest heavily in the Seventh Street corridor, choosing it as one of three neighborhoods—along with East Palo Alto and San Jose's Mayfair—in which to implant a Neighborhood Improvement Initiative.

Lacking a local apparatus, they chose the San Francisco Foundation to help them administer the program, and selected a young man named Fred Blackwell to coordinate things on the ground. Blackwell arrived in West Oakland with a rich set of connections. He'd grown up in East Oakland, the son of Angela Glover Blackwell, a star of the nonprofit world, and his uncle David Glover had been a leader in West Oakland for years. He had easy access to the old guard like Paul Cobb and Ellen Wyrick-Parkinson, but there was still a lot to learn.

As he got his feet wet, a small woman appeared, at least a foot shorter than him. This was Margaret Gordon. "Margaret was one of the first people who was like, let me teach you West Oakland," said Blackwell, now the CEO of the San Francisco Foundation.

As it turned out, they had a person in common: Espanola Jackson, a longtime community organizer in Hunters Point, sometimes included in that area's Big 5. Margaret struck Blackwell as the Jackson of the local area. Blackwell spent hours talking with, fighting with, and learning from Margaret.

Hewlett's initiative encompassed more than half a dozen neighborhoods in West Oakland all with their own histories, idiosyncrasies, and demographic realities. Acorn had mostly very-low-income residents in public housing, led by Janet Patterson and their neighborhood association, whereas Oak Center was filled with middle-class Black people whom Ellen Wyrick-Parkinson had been organizing for decades to protect their neighborhood from local government and local residents alike. Many others, like Monsa Nitoto, a local artist and activist, had earned the right to play a major role in community deliberations.

West Oakland was undergoing tremendous demographic change as Latin and Asian immigrants moved in. White people, who had abandoned West Oakland for a generation, were coming back. "West Oakland as a community was the canary in the coal mine for what we're seeing today around the gentrification of this place," Blackwell said. "The artists were moving in. There were these live-work situations that were popping up. Speculators were walking around the community asking people who were seniors if they were ready to sell their homes." More and more Black people were moving out to the industrial suburbs of the Bay Area and to Sacramento and Fresno. Property began to sell for more and more money in absolute and relative terms. Between 1996–1997 and 2001–2002, home prices in West Oakland rose from just 30 percent of the Alameda County average to 60 percent.

With all the changes, when the Hewlett planning kickoff meeting occurred on November 14, 1998, 350 people showed up. Given the history of the area, the people did not welcome Hewlett with open arms.

"What were the things that people were carrying?" Blackwell asked, remembering back, ticking through the community's issues. "The construction of the post office, the aboveground BART rail and the BART station itself, the freeway building through that community. So there was a big part of it that was about government just riding roughshod at multiple levels through this community and doing what it wanted to do, damn what the community said, needed, or wanted . . . They were also carrying the fact that post–Loma Prieta, they [Caltrans] tried to rebuild the freeway in the exact same place and the community stood up and won."

At the same time, the community had its own ideas about what West Oakland was about. "It was carrying the legacy of the Black Panther Party and its history. It was carrying what the community felt like were wins at the Eighteenth and Adeline intersection," he said. "So you had the library on one corner. You got DeFremery Park on another corner. You had the senior center, which was the result of a lot of organizing, on another corner."

This was a place where the more politically involved residents like Margaret had been involved in three, four, five major planning and organizing processes. The decommissioning of the Oakland Army Base, the freeway, the port, redevelopment stuff. People thought they knew how the city worked, and their idea of it was mostly adversarial. No matter what anyone told them about good intentions, they could point to . . . well, the whole of history of the neighborhood as a kind of refutation.

Given all that, Hewlett's idea—as implemented by Blackwell and the San Francisco Foundation—was that the process should be *resident-driven*. "The most important feature of this initiative is its emphasis on residents having a real say in determining how and where resources will be distributed," Blackwell wrote in an email to local organizations. They would come up with a plan that would send money into the places in the community where local residents actually wanted it. It seemed as if, *this time*, it could be different.

It did not turn out to be different. The Hewlett initiative ran into problems from the beginning. Though it was supposed to be resident-driven, no one *really* knew what that would mean. Residents poured in effort, volunteering 3,000 hours in the planning and implementation of the project.

The foundations had their own interests. The board brought in one executive director, Allen Edson, whom Hewlett deemed inadequate. There were grant-making issues, both dispersing them too broadly and giving them only to organizations whose employees joined the initiative.

The ignominious end came when a new resident board was voted in and fired Edson, but then the board, the San Francisco Foundation, and Hewlett got locked in a stalemate over picking a replacement. Some wanted a young Black firebrand, Desley Brooks. Others (you can imagine who) did not. The situation devolved and Hewlett simply picked up and left, not even halfway through its vaunted seven-year commitment to the area. Hewlett, in its final analysis, primarily blamed residents themselves for "contentious behavior" and focusing on their own "set of narrow interests."

For all the hemming and hawing in the Hewlett world over how the money was spent, the grand total of $1.6 million in grants that the foundation put into West Oakland was a teensy-tiny fraction of a single gift that Hewlett gave to Stanford in May 2001. In general support of the humanities at the school, the Hewlett Foundation donated $400 million, the single largest gift made to any university to that date. The message was clear to people in West Oakland: *they* had to scrape and claw for a few hundred thousand dollars with all kinds of strings attached, while the already privileged on the Peninsula got a check *250 times as large* to do with what they pleased.

All that said, there were some good things that came out of the Hewlett Initiative. None became more lasting or significant than the Pacific Institute's Environmental Indicators Project. Hewlett had funded the organization to bring a group of interested community members together to develop a common understanding of the neighborhood's priorities and problems under the rubric of "neighborhood health."

Over seven long meetings, they made lists of West Oakland's problems, placing them on sticky notes. The staffers would then cluster those problems. As the process went on, they narrowed the many suggestions down into a short list of seventeen indicators of neighborhood health that captured the key concerns of the residents. Then the researchers went off and assembled as much data as they could from government agencies and private companies.

Eventually, they formalized their data into a report, *Neighborhood*

Knowledge for Change. While the indicators ranged from voting power to bike lanes, the most obvious cluster of problems highlighted air quality, toxic exposure, *and* the displacement of residents. From the beginning, the Environmental Indicators Project construed environmental justice as a holistic approach to neighborhood health. It wasn't just about *reducing* air pollution for anyone but for the *people who actually lived there.*

That meant they would tackle all kinds of problems. Working with other environmental groups, they got a yeast plant shut down that had been pouring disgusting carcinogenic pollution into the neighborhood for decades. They'd get involved in the push for a co-op grocery store across from the BART station. They'd follow through on the construction of Mandela Parkway. They'd help create a truck route that reduced the trips made by semis near schools and homes.

As the organization took shape, Margaret and her co-director, Brian Beveridge, took on more responsibility, conducting what scholars in public health were calling "community-based participatory research." They focused on collecting their own data, understanding the ultralocal conditions, *and* playing the broader politics. "I had to learn how to be a nerd, dealing with all this data," Margaret joked.

In downtown, Phil Tagami was focused on a different kind of development. The building that became Phil Tagami's home base was once an old department store, right across a plaza from City Hall. It was the type of building that once inspired civic pride, and inside, you could glimpse an earlier Oakland's faded glory. Though it had fallen into disrepair, and been damaged by the Loma Prieta quake, a massive windowed dome capped its large open atrium. They don't build buildings like that anymore. The bones were good, but who would pay to restore that kind of building in a place like Oakland?

Every couple years, a California newspaper would run a halfhearted revitalization story, but it always included quotes like "Oakland sees itself as the not-so-beautiful contestant in a contest," or "We're getting to the point where it's either cappuccino or crack," from Oakland citizens.

For decades, until the late 1990s, new office buildings simply stopped being built. Why build in complex Oakland when the green-field suburbs of south Alameda County and east Contra Costa County were booming? Beginning in 1981, Bishop Ranch, a patch of land east of Oakland off the 680 freeway, added 10 million square feet of office space. That was substantially more than the *total* inventory of Class A office space in downtown Oakland even in 2024. No one was building housing in Oakland, either, because few people were building anything. "It's so frustrating to have people from elsewhere always feeling sorry for you about where you live," one native told a newspaper reporter. "You can hear the pity in their voice."

There were some developers interested in that old building, including some big boys like Shorenstein. But among those bidding for the project, there were some locals, too, namely the California Capital and Investment Group (CCIG).

The company had come into being when the owner of an athletic club named Len Epstein took a shine to two young, hardworking employees: Phil Tagami and Mark McClure. Together, they looked around Oakland for business opportunities and eventually settled on the idea that they'd buy distressed buildings and fix them up. They did their first deal over near Lake Merritt, transforming an old building into an athletic club.

Tagami became the public face of the company, and an important player in local politics. His work volunteering for Chappell Hayes had brought him into contact with David Glover, a member of the powerful Black family who was then the head of OCCUR, a redevelopment effort that had grown out of urban renewal reforms. These community leaders taught him the ropes of working with local governments.

As he burrowed into the power structure, he found new friends. Here was a guy willing and able to invest himself in the city, and who was *a workhorse.* Tagami assiduously cultivated many different political connections. When Tagami was twenty-eight, Mayor Elihu Harris of

Oakland declared May 25, 1993, Phillip H. Tagami Day. Tagami had met the future councilman Ignacio de la Fuente volunteering with Hayes's city council campaign, and they built a deep relationship. He got close to State Senator Don Perata and U.S. Representative Barbara Lee. When Jerry Brown moved to town, there was Tagami, happy to show him around and cut him in on deals.

Tagami, by all accounts, was simply relentless in his drive to rebuild downtown Oakland. Would that make him rich? Definitely. But even his harshest critics don't doubt that he wanted Oakland to get better. As the '90s wore on, Harris succeeded in getting some investments into downtown Oakland. A few office buildings were built. An old market renovated. The city even purchased the grand old Fox Theater for $3 million. Tagami approached Oakland about restoring it and, in his own words, "felt laughed at."

Even that old department store building seemed out of reach. The city planners and local bureaucrats wanted the deal to go to an established national entity. But they didn't realize just how much political capital Tagami and his partners had accumulated. "After all these people had their little speech about how we should have the Shorensteins of the world do this, the mayor literally stood up, slammed (his hand) on the table, and said, 'This isn't your decision,'" Tagami's partner Epstein recalled. CCIG got the deal, including a $12 million loan from the city. Tagami assembled the rest, and by 2001, they'd poured $50 million into the building and reopened it to much fanfare.

During those years, if you were working late or needed to talk business with someone, there might not be a single place you could meet clients in downtown Oakland. So, down in the basement, Phil installed a wine cellar and sitting room. You walk in off a fluorescent-lit service hallway into something like a miniature private club. Power in Oakland often liked to work underground.

18

Life, Death, and the Port

As Margaret and Brian were getting the West Oakland Environmental Indicators Project set up independently from the Pacific Institute, a new public health director arrived in Alameda County.

Tony Iton, a brilliant Black doctor, came from Quebec. He'd gotten his MD at Johns Hopkins in the 1980s, in the midst of the crack crisis, as Reagan's cuts to the social safety net opened up new holes for people to fall through. What he saw all around him was "medical apartheid," a system that forced him to treat, as he put it, "social ills with pills."

Baltimore had been hit by a very similar constellation of problems as West Oakland. "One was crack cocaine. Two was HIV/AIDS. Three was violence. And four was mass incarceration and militarized policing," Iton recalled. "All of those things descended on Baltimore." On the streets there, Iton saw despair, a shortening of the horizon for everyone exposed to ghetto environmental conditions. He coined a term to describe the lack of agency and belonging that he saw in the streets: *futurelessness.*

It drove him to try to figure out the broader systems that sent broken children to the hospital. He spent some time on Capitol Hill and then headed to Berkeley, where he added a master's degree in public health and a law degree to his credentials. His first job in government was as the public health officer in Stamford, Connecticut, and then he came west in the early 2000s.

There was a grisly aspect to the job. For every person who died,

Iton signed his name to a paper confirming the death. And as he put a pen to each certificate, he began to consider what kind of dataset the health department happened to possess. For every death in Alameda County, it had a record of the person's age, cause of death, race and ethnicity—and where they lived. Ultimately, Iton used 400,000 death certificates from 1960 to 2005 to pull together a stunning report, which was made into a documentary of the same name, *Unnatural Causes*. The biggest fact that emerged from the report was that the life expectancy of a Black person born in West Oakland was *fifteen years shorter* than that of a white person born in the Oakland Hills. That's similar to the life expectancy difference between Sweden and India.

The outcomes did not result primarily from gang violence or HIV. People in West Oakland were dying from the same things that kill most people—cardiovascular disease, cancer, etc.—but they were dying sooner. And the gap in life expectancy wasn't narrowing but actually had been expanding since 1960, and doing so rapidly since 1980.

The life expectancy differences were not *only* rooted in race. Place and wealth were deeply significant as well. A Black person living in the hills had a life expectancy of almost seventy-seven years, longer than that of a white person living in the flatlands. Across the racial and ethnic board, immigrants lived longer than those born in the United States, too. But taken as a whole, something terrible was happening in America, and it had a lot to do with your zip code.

Iton argued that poverty and racism forced people into deeply stressful environments. It wasn't *one* thing but the cumulative impact of all the things. The air, the noise, the crime, the grime—all the stuff that surrounded Margaret when her kids were little and getting sick all the time, that were all around her still in West Oakland. All this stuff—whatever the actual biophysical mechanism—irrefutably kills people earlier. Sometimes, people call it "weathering."

The report his department produced is a tour de force about not what *health care* should look like but what kind of society would generate healthy people. They boiled it down to ten simple sentences for their television documentary.

Health is more than health care.

Health is tied to the distribution of resources.

Racism imposes an added burden.

The choices we make are shaped by the choices we have.

High demand + low control = chronic stress.

Chronic stress can be deadly.

Inequality—economic and political—is bad for our health.

Social policy is health policy.

Health inequities are not natural.

We all pay the price for poor health.

Firing a broadside about how economic inequality was literally killing people might seem surprising from such a pedigreed doctor. But Iton had a more radical background than you might guess. He was married to a Salvadoran woman whose father had been a leftist guerrilla commander. A socialist, she'd come to the States and ended up working as a community organizer for the Service Employees International Union and other labor organizations. During her campaigns, sometimes she'd have Iton go out with her into the community to talk about public health. "I'd come into these conversations and think, 'Oh my God, this is what public health should look like,'" he recalled. "Organized community members coming together demanding change that will protect the well-being of both them and their community. That's public health! And so it clicked in my head like: Where's power building in this work? Traditional public health comes at it as a technocratic enterprise. And I realized that this was a war between technocracy and democracy, and we needed democratic solutions as opposed to technocratic solutions."

That steeled him in the initial research he'd done into the life expectancies of different areas of the county. While mapping of diseases and deaths was a storied part of public health history and practice, no one wanted to deal with the implications of these statistics, even if they had the data. The city of Philadelphia was using death-certificate data to show how mortality rates varied with poverty rates. "We called them up and said, 'Would you map this?' They said, 'No, no, no, no, no, no.

It's politically contentious,'" Iton remembered. "But that's exactly why you'd do it."

His own department staff resisted doing the analysis he'd asked for. A senior epidemiologist objected that they couldn't examine small geographical units by compounding deaths over three years to get sufficient sample sizes. First, she threatened to quit. Then she called the National Center for Health Statistics, part of the CDC, who then sent him a letter saying he could not use the death-certificate data to create the maps and comparisons he wanted to. He pushed back hard enough to get the go-ahead.

So they did it for Alameda County. Then they did it for Baltimore and places in Ohio, all over the country. And everywhere they could show a map like that, it became front-page news. "To me, that was the point," Iton said. "It was to have this public conversation about how we're creating socioecological conditions equivalent to war-torn countries in our cities. And we blindly suggested that this was natural. This is just the way it is."

In West Oakland, residents' health was always tied to the port trucks, their engines, the fossil fuels that powered them, and the pollutants that poured out of their exhaust pipes and into the lungs of residents.

For decades, the cost of this air pollution was simply placed onto West Oakland residents. They were poor and Black, politically disempowered, and suffering from the compounding effects of environmental racism.

Shipowners made money. Terminal operators made money. The port itself made money. The cargo owners, such as Target, made money. Even the ILWU took its share for longshore workers. Unfortunately, the battle to clean up West Oakland's air didn't pit community residents against the real power structures that were causing one in five children in the neighborhood to suffer from asthma. Instead, the fight pitted poor, mostly Black residents against poor, mostly immigrant men.

Moving containers in and out of the port is called drayage, and it bears little resemblance to the long-distance trucking you might be familiar with. Drayage truckers make short little trips to regional ware-houses. And they were barely scraping by. In one survey published in 2007, more than 90 percent of the truck drivers were born outside the United States, and the median driver made less than $11 an hour. Two-thirds did not have health insurance, and they worked an average of eleven hours per day.

Trucking had been a regulated industry, largely unionized by the Teamsters, but in the 1980s, it was deregulated and shipping rates plum-meted. The union was decimated. The drayage industry splintered. A hundred different firms were running into and out of the Port of Oak-land, all technically "independent contractors," even if they mostly or exclusively worked for one big corporation. Large importers like Tar-get, Walmart, and Costco (among many others) could use their scale and market power to squeeze all these tiny firms, making it nearly impossible for the little guys to make any money. The truckers spent hours waiting for boxes sometimes, a factor largely out of their control, and they just had to eat that time/cost, rather than being compensated for it by the big companies they were driving boxes for.

To put it bluntly: drayage became the worst job in trucking, an in-dustry that itself had become much, much worse. And, if you squinted just a tiny bit, the port truck drivers looked like . . . gig workers? "The American work force as a whole is starting to look more and more like the port trucking industry," one trucking scholar observed. "Risk is be-ing shifted from companies to workers throughout the economy."

With such thin margins, the drayage truckers were always looking to cut costs. Sometimes that meant not parking in their designated, paid lots out by the port, and instead parking on the streets of West Oakland. Of-ten, it meant purchasing used trucks, older equipment that was no longer suitable for long-haul trucking but that worked for their short trips. This helped financially, but it also meant the oldest, dirtiest diesel engines in the country were the ones rolling through the streets of West Oakland.

The drayage truckers were not villains. "We came to America for a better life . . . but that's not happening. Instead we are treated like disposable workers. Truckers are just like everyone else," said one port trucker, Kulwinder Singh. "We want good jobs, and a better future for our families." The truck drivers who plied the roads were essential . . . and also disposable.

Starting in the early 2000s, Margaret and the coalition of people growing around WOEIP had begun to work with the truckers. They spent two years working with trucking associations and the Port of Oakland to create a truck route that nudged the trucks away from key residential areas, where their diesel emissions were more likely to harm people.

It was a huge, boring achievement, and Margaret recalled it as perhaps the most difficult organizing experience of her life. Both the residents and the truckers just had so many legitimate grievances about their lives that it was hard for anyone to give ground. Even after they worked it out, the enforcement was spotty, and major problems remained.

Meanwhile, in the mid-2000s, the Teamsters had begun to organize drayage drivers up and down the West Coast. They created a group called the Coalition for Clean and Safe Ports, which in Oakland teamed up with EBASE, the East Bay Alliance for a Sustainable Economy. They put out a damning report, "Taking the Low Road," about trucking at the Port of Oakland, which showed just how exploited drivers were. Worse, because they were classified as contractors, anti-collusion laws prevented them from taking any serious collective action. To labor people, sectoral bargaining made so much sense for these particular workers, just as it had for the dockworkers.

The innovation in the EBASE report was tying these terrible working conditions to the environmental problems in West Oakland—and offering a solution that became known as the "concession model." The plan first got traction in Los Angeles. It would require that truckers working in the port be declared actual employees, not "contractors." The resulting bigger trucking companies would be able to bargain more evenly with the large shippers and retailers who brought goods through the port.

EBASE brought in Tony Iton, the Alameda County health officer, to write a foreword for the report. "It is abundantly clear that the current port trucking system is contributing directly to an acute social and public health crisis in the neighborhoods adjacent to the Port as well as among the port truck drivers themselves," Iton wrote, before endorsing the concession model.

The Teamsters came on strong. They wanted to prove out this new model for organizing drayage drivers. "We had been trying to figure out how to organize the ports for twenty-five years," Doug Bloch, who ran the Teamsters' efforts around the Port of Oakland, told me. "It was the first place we lost with this independent contractor model. We had done everything from co-ops to militant worker action." None of it had worked until they'd hit on the concession model. Margaret, for her part, was thanked in the EBASE report and spoke at a Coalition for Clean and Safe Ports event.

But Margaret had other loyalties, too.

She'd been immersed in truckers' problems for years. As an environmental leader, she might have been an odd person to make friends with the truckers, who, after all, were the ones driving the rigs creating so much diesel pollution. But Margaret saw in them other oppressed people. She knew they, too, breathed the same exhaust. They lived in the neighborhood or commuted from other poor places. She had made allies in the local community as they worked through the brutal process of threading the truck route around the neighborhood.

One of those guys was Bill Aboudi, a Palestinian American who ran a trucking company and provided truck parking out on the old Army base. And he absolutely did not want to deal with the Teamsters or a bunch of employees. He fought them tooth and nail. The Teamsters didn't mess around, either. With other environmental allies and EBASE, they hit him with a lawsuit, showing he had never applied for stormwater permits and had dumped nasty stuff directly into drains that went out into the bay. (The parties settled the lawsuit.) Separately, Aboudi lost a million-dollar class-action lawsuit, with a court holding that his trucking company had encouraged people not to take breaks. And he got a job for one of Margaret's cousins. To the Teamsters, he was everything wrong with port trucking, a low-level crooked businessman out to screw the people two notches down from him on the economic ladder.

So, when the Teamsters began their big drive around the port, Margaret knew there might be trouble. This was a cut-and-dried labor organizing situation, from their perspective. Only with an effective union covering the whole of the port bargaining on their behalf could the truckers actually come out better than they'd gone in.

The bargaining chip to get everyone on board was that the truckers *needed* to upgrade their trucks in order to clean up the West Oakland air. State law said so. They would need collective power to pass the cost of those upgrades on to retailers and shipping lines in the form of higher rates. Otherwise, Oakland's air might get cleaner, but the truckers would be worse off.

But some independent truck drivers maintained that they didn't *want* a union. They had their own trade associations and representatives. To the Teamsters, a bunch of small-time individual truckers were getting screwed, even if they couldn't admit it.

But as Margaret encountered them, the big union trucking shops didn't hire the local people. The local trucking companies were the ones that actually ended up providing work for the people of West Oakland. And she was a loyal person. These independent truckers had stuck with her.

The conflict between Margaret and the union organizers never came fully to a head or to light before the Supreme Court ruled against the Teamsters' port-organizing model. But the Teamsters would not forget that Margaret didn't get on board.

By 2007, through hundreds and hundreds of community meetings and her environmental justice work, Margaret had become one of the most respected community leaders in Oakland. And if she had enemies, she also had friends.

Ron Dellums, the former congressman, became mayor of Oakland that year, succeeding Jerry Brown. His community contacts convinced him to appoint her to the Port Commission. No one from West Oakland had ever been a commissioner. Whatever other concessions to liberal politics and populism that the succession of mayors had made, none of them had been crazy and/or committed enough to put someone who might possibly gum up the wheels of the economic engine that everyone believed the port to be. For example, Jerry Brown had appointed Phil Tagami to the Port Commission. A developer on the commission made sense to the powers that be. Margaret? Not so much.

"I'll be bringing the experience of a community that other commissioners don't see, they don't smell, they don't hear," Margaret told a reporter. "I can tell them about the trucks riding up on the sidewalks in West Oakland and cracking the pavement where residents have to walk, or idling under people's bedroom windows, or the dust and dirt and soot they leave behind them on the sides of houses, or the smoke billowing out of ships that comes over into our neighborhood, or the mothers sitting up with their kids hacking and coughing all night, or the people missing work because of respiratory problems."

Yes, she could tell them all about that, but did the city want those problems to be brought out into the open by a port commissioner? They sure hadn't wanted such knowledge front and center for the *first fifty years* of the port's container era. In fact, hiding the environmental and public health costs to West Oakland of the operations was central to the political project of the port.

By the mid-2000s, the port was $1.4 billion in debt and was paying more than $100 million a year to service it. It was desperate to keep up with the ports in Southern California and Washington State that had been eating into Oakland's market share for twenty years. In that kind of competitive environment, the idea of putting someone on the commission who might not *only* care about the economic growth of the port was scary to the city's elites in business *and* labor.

Her first nomination confirmation was postponed when Mayor Dellums realized he didn't have the votes. Phil Tagami's old ally Ignacio de la Fuente was the center of the opposition.

But Margaret's people kept pushing. U.S. Representative Barbara Lee personally called at least one city councilmember, Chappell Hayes's widow, Nancy Nadel, who had known Margaret for years. Councilman Larry Reid had, too, though he was not sure about her nomination.

"I've known Margaret for a very, very long time," Reid said in a city council meeting. "Her family is close to my family. And even when she heard that I may not be supporting her, she just laughed with her sense of humor, 'That's OK, Larry, I understand, but I still love you.'"

In the end, every city councilperson voted to confirm her except Desley Brooks, by then a councilwoman from East Oakland. If she felt any pull for Margaret as a fellow Black woman, she did not reveal it in her remarks at the council meeting. "The port is getting ready to face some very tight times. We don't need a commissioner who can only speak to the health and environmental issues," Brooks said. "They also need to be able to be visionaries and take the port to the next level. I told the mayor I hoped he would prove me wrong in this appointment."

In the end, Margaret was a compromise. For business interests, they could say, well, at least she wasn't a union person. For progressives, she might not be a union person, but she had a rock-solid reputation as a community activist, and she'd spent the better part of a decade learning the ins and outs of air quality management. After so many years of fighting from the outside, Ms. Margaret found herself on the inside, a business card–carrying member of the leadership of the city.

In practice, though, would people want it pointed out—over and over—that the economic growth of the port meant sacrificing years of Black lives? That truth was uncomfortably close to the premise of America itself.

Her own hard-won knowledge of air quality monitoring, environmental justice, and port operations was one thing, but the ins and outs of port business were another, and Margaret knew it. "How can I be on par with these other people who were executives in banks? Or they were lawyers. Or they owned businesses or were accountants. How do I be on par with them and be able to speak the same language?" she said. "And at the same time navigate myself to a place where I could talk about the injustice?" She forced herself to do every bit of reading that they gave out, turning page after page in thick binders.

Before she spoke at any meeting, she would write down her three key points, and then stick to them. There was no way anybody was going to catch her meandering or not sticking to the issue at hand. At the same time, she realized that the people on the Port of Oakland board didn't always have Oakland's best interests in mind, to say nothing of West Oakland. Part of their privilege was assuming that what was good for the port's development would be good for everyone. She found herself sticking up not just for local *environmental* issues but local issues of all kinds. These people all wanted to see the global picture, the Pacific Circuit, not West Oakland.

Margaret had her principles. What mattered most to her were her neighbors. The more ideological issues of labor and the environment, of capitalism, were secondary. The language of the white-collar meetings began to flow from her tongue. "Now, I go in there and I say, 'This doesn't reflect the needs of our community,'" she said. "And then I get outside the door and I say, 'Fuck that motherfucker.'"

What was happening along Seventh Street was abstract to the other people. They saw the spreadsheet, not the community. "I live in a place where I'm gonna see somebody. You're not gonna see the same people I see," she would tell the other commissioners. "I'm gonna see somebody's mama, grandmother, uncle, cousin, sister, brother. That know me and are going to ask me why I voted this way."

Margaret's crowning achievement as a port commissioner was the Maritime Air Quality Improvement Plan in 2009. It was the master document containing a quiver of strategies for actually cleaning up the air in West Oakland. Upgrade the trucks, reduce overall trips, etc., etc. "It's always been the trucks," she said. "Old, dirty trucks everywhere."

She built connections with other agencies, especially the Bay Area Air Quality Management District, and sometimes she played them against each other. One time, when the port was dragging its feet, she went to the state capitol with Bay Area Air Quality staff. "We had two of my grandchildren and two of my godchildren, saying, 'We can't wait around to get lung transplants,'" Margaret said. "I had them with their asthma machines and their inhalers." The port officials weren't happy, but the stunt worked. She laughed. "I embarrassed the shit out of them."

In the end, they got the measures enshrined at the port and Margaret made sure they were implemented. The port itself cited the plan for an incredible 98 percent reduction in diesel particulate emissions from trucks in West Oakland by 2015. "This is a story about people coming together to fight for justice. And they won. And they won big," said Tony Iton, who saw the battles up close. "And they won in a way that had influence on what's happening in ports all over this country. And in fact ports all around the world. I think those are the true heroes, the Margaret Gordons . . . They drove this change."

Sometimes, though, the stress of it drove Margaret crazy. "If I got too pissed off about the conversation going on, I would get up, take my purse, go inside the bathroom, scream inside the purse, AHHH-HHHHH," she said. "My nose would be sweating, back of my head would be wet—I was going through menopause—then I would drink me some ice water, some tea, and then go back into the room."

19

The Revolution Cafe

There's a story that almost everyone alive in the 2000s knows about mortgage-backed securities, collateralized debt obligations, and people of color. It goes like this. Once the pool of traditional *creditworthy* people was exhausted, the people selling mortgages began to seek out anyone else who might want a loan, which included large numbers of people of color. With financial instruments that purported to spread and dilute risk, suddenly money was flowing to everyone, even people traditionally excluded from the mortgage finance system. When the housing boom went bust, many of these people could not make their payments and we went through the financial crisis, because it turned out the entire global economy had come to depend on some of the most oppressed people in America's ability to pay their bills. So much Black wealth got wiped out during the housing bust that the Black-white wealth gap grew to be as large as it had been in . . . 1968.

To understand what really happened, maybe it is best to see the system of mortgage finance in America from the vantage point of the peak. The number and volume of loans made in the early 2000s was astonishing. Huge swaths of the *global* economy were dependent on securities that had been created out of the debt owed by tens of millions of Americans on their homes. But this wasn't necessarily a bad thing! On the contrary, the mortgage payments of everyday people were nearly as good as cash for investors, and in return, the financing available to just about anyone to buy a house was easy and cheap. Most important for our story, after decades of disinvestment in Black neighborhoods, the money taps were *flowing*.

From 1999 to 2005, the amount of credit extended in West Oakland went up by more than 8,500 percent! The $243 million that flowed into West Oakland in 2005 was 1.5 times the total credit secured by West Oaklanders in all of the 1990s. That's big money. And, for once, in this bonanza, Black people were getting a substantial chunk of the dollars. At a quantitative level, it looked like a triumph.

The easier money trickled down to people buying houses, refinancing their mortgages, or pulling equity out of their homes, which had been increasing in paper value for years and years. The cost of borrowing kept dropping as more dollars became available for mortgage financing. Millions of Americans made good use of the Chinese export profits, the Middle Eastern oil money, and the German central bank investments that filled the pool of money on which they drew.

While some people began to worry, others did not. Traditional ways of thinking about risk had held back extending mortgage credit in ways that were clearly prejudiced against, specifically, Black people, but also many other minority groups, as we were called at the time.

And it wasn't *not* true. The banking system *had* been local, racist, and conservative, with relatively limited access to capital from faraway sources. New technology allowed everyone in the financial network to imagine that they'd escaped the previous rules of risk, many of which *were* bad. The evidence was all around West Oakland. People were getting loans left and right. People who'd *never* been able to get a loan in previous years could buy houses.

The money went looking for the people, *not* the other way around. Billions of dollars' worth of containers filled with imports came to the United States. Dollars went back to Asia. Most Asian states decided to invest those funds *back into* the United States rather than spending the money at home. Buying your crap at Target, then, helped lead to the flood of cheap mortgage credit.

The increasingly complex mortgage-backed security was the perfect financial product for investors, in part because it could mask change in the underlying assets. Michael Osinski was a Wall Street programmer who helped create the software that manufactured mortgage-backed securities. Osinski had an older colleague who knew just how to describe the magic of the MBS: "You put chicken into the grinder and out comes sirloin."

Who would believe chicken could become sirloin? In fact, that is one of the specialties of the technology industry. Old limits and old rules could be disrupted! Chicken can be sirloin! Osinski wrote a first-hand account of his time in finance, and it's one of the few that attempts to reckon with the role tech played in the subprime crisis. "The first collateralized mortgage obligation, or CMO, was created in 1983 by First Boston and Salomon Brothers, but it would be years before computer technology advanced sufficiently to allow the practice to become widespread," Osinski wrote in *New York* magazine. "Massive databases were required to track every mortgage in the country. You needed models to create the intricate network of bonds based on the homeowners' payments, models to predict prepayment rates, and models to predict defaults."

These models served two purposes: they did the core work of creating CMOs *and* they obfuscated the chicken.

A mortgage was one thing. A savings and loan would have taken the money in from its customers and lent back out some piece of it, making money from the payments of the person who took out the loan. But Osinski's software, and that being developed at other financial institutions, took the *idea* of the underlying assets and transformed it into projected cash flows that could be purchased however an investor might like. One could take the safest layers (or tranches, as we learned to call them) and get paid a little less or take on more risk and get a higher premium.

The subprime mortgage providers could believe they were driving "a time of innovation." The mortgage-backed securities creators could imagine that their algorithms had cracked the code on spreading and limiting risk. And all these automated systems, the scoring methods, gave almost everyone the sense that someone, or *something* at least, had figured everything out. The former Wall Street quant Cathy O'Neil described in her book *Weapons of Math Destruction* the faith people had.

They assumed that "crack mathematicians in all of these companies were crunching the numbers and ever so carefully balancing the risk," O'Neil wrote.

Working with past loan performance data, modelers at banks and at the credit-rating agencies were able to show that these investments had been quite safe, and declared them likely to remain so in the future. In 2003, only 10 percent of the mortgages made in America were considered "subprime." Those numbers ballooned for the next several years.

Then came the cooling of the housing market. West Oakland's story is typical. After the median sale price of a single-family home topped $500,000 in 2006, the bottom fell out. Prices fell to around $200,000 by mid-2008. People who expected to be able to flip their homes could not. Small-time real estate investors got caught in bad positions. Loans with short-term teaser rates reset higher. Balloon payments came due. The foreclosures began in earnest as people began to default on their loans.

Investors in collateralized debt obligations thought they were buying *good-as-money* investments that just so happened to generate higher returns. Moody's and other rating agencies had given them their blessing. People expected them to perform like older mortgage-backed securities, in which the foreclosure rate was tiny, maybe 2 percent. But instead the ultra-high-risk mortgages included in these portfolios performed terribly, as some people defaulted before sending off a single payment. These were not AAA investments. They were junk investments disguised by complexity and the sheen of novel technology. As the housing market soured, the assets became *toxic* because they were both worthless and corrosive, eating at the very foundations of the global financial system.

The whole structure began to collapse. As different parts of the financial ecosystem began to realize what was going on through 2007, the implosion of the U.S. mortgage market spread across the world. The toxic debt threatened to destroy Wall Street, with unknown consequences for Main Street. Lenders pulled their money off the table. The credit market was frozen.

The Bush and Obama administrations engineered a bailout of epic proportions, but unlike the New Deal programs, which had dealt more directly with homeowners, these fixes put money in the hands of banks and asked them to cut deals with homeowners. They mostly did not keep people in their homes. The incentives in the bailout often made foreclosure the most profitable path for the banks.

And in a sure-to-gentrify neighborhood ten minutes from downtown San Francisco, why wouldn't a bank want to take possession of a house? They did, and not just in West Oakland. Across the city, there were more than 10,000 foreclosures, and the crisis really didn't end until 2010. New kinds of companies began buying up foreclosed single-family homes. One Urban Strategies Council analysis found that investors managed to lay their hands on more than 40 percent of the foreclosures in Oakland—and more than 90 percent of those were in the flatlands. At the same time, other companies began to string together massive networks of rentals, mostly from the foreclosed housing stock. Then these investors could take out liens against these homes, popping their money free to buy more homes. Those liens got rolled into big debt balls again, which were sold to investors.

Looking at the mortgage data in West Oakland after the crisis, we see a total collapse of credit for Black people. From 2008 to 2018, Black people in the neighborhood received a total of 370 loans, fewer than half the loans that they'd gotten back in the 1990s. People's credit was destroyed. While several financial institutions, such as Countrywide and Wells Fargo, settled lawsuits for making predatory loans, the vast, vast majority of people who made money from the former ghettos of America got away with it, just like the people who polluted West Oakland land took their money and left the toxic chemicals. As the protest chant went, "Banks got bailed out. We got sold out."

For white people in West Oakland, the boom days of the 2000s quieted down, too, but not nearly as much as for other racial groups. Through the early 2010s, white people were receiving 60 to 70 percent of the total dollars flowing into the neighborhood, despite representing less than 15 percent of the population. There is still a large Black community in West Oakland, but the white people they live alongside live in a different financial world, with economic gaps as large as they were before the civil rights movement.

There simply is no way to describe how catastrophic the subprime boom and bust was for Black Americans, a fact that must not be forgotten in the other conversations we have about how American cities are changing. On the surface, Oakland bounced back from the subprime crisis, but the people who were living there at that time did not. They just now live in the industrial suburbs.

For those who have managed to stay, even the composition of the rental market has been transformed. Oakland birthed one of the first major corporate landlords, Waypoint, which combined with a passel of other similar companies into the mega-corporate landlord Invitation Homes, which controls 80,000 houses, is traded on the New York Stock Exchange, and has a market value of more than $20 billion. During one of the transactions, an early Waypoint investor bragged, "We are gratified in playing a role in turning a cottage industry into a permanent asset class within the $30 trillion U.S. housing market."

In West Oakland, the biggest buyer of homes was REO Homes, the investment vehicle of Neil Sullivan. Sullivan has amassed hundreds of properties, most of them in West Oakland, and served hundreds of eviction notices to residents. It's a more local operation than Invitation, but it is equally connected to the circuits of global capitalism. The company's largest backer has been the billionaire hedge funder, climate hawk, and presidential candidate Tom Steyer, along with his friends.

Across the country, private equity firms were purchasing 15 percent of single-family homes by the end of 2021. From 2012 to 2019, rents across the ten highest-poverty zip codes in the Bay Area went up 67 percent, more than double those of wealthier neighborhoods. In West Oakland's 94607, where Ms. Margaret lives, rents went up 80 percent.

Given the structural conditions of the nation and the Bay, as they were developed by racial capitalism, West Oakland was being emptied of Black people. The people who are replacing them will be much more likely to work in the technology industry and be white and Asian. In

San Francisco, one data-driven real estate broker discovered that one out of every two identifiable purchasers—50 percent—were in the tech industry. And West Oakland is only a ten-minute BART ride away.

←--- --→

Sometimes, riding my bike around Oakland, dodging potholes, bumping over abandoned railroad tracks, through the shadows of new towers, under leafy trees, past the encampments, through the cold mornings, I try to imagine the money flowing in and through the place itself. In high school, there were those diagrams in physics that would lay out all the different forces acting on an object. What if you could see that, but for the economic pressures on a place? Imagine where each dollar came from, where it sat in a server farm pretending to be something like gold, and trace the path it would follow once it left one computer and went to another, or moved as paper from one hand to the next.

There was only one coffee shop on the west end of Seventh Street in the 2010s. Sometimes it seemed to be called the Seventh Street Cafe, other times the Revolution Cafe. If it was open, jazz was usually playing, and there was old Panthers paraphernalia on the walls, a Malcolm X poster, an Aztec god, the symbol for the artist formerly known as Prince. The cafe was attached to a big open lot filled with old furniture and falling-down little improvised structures. The site of the former Lincoln Theater, it was a good place for a party, but for anyone passing by on the BART train overhead, it might have been a little hard to parse what was going on.

The place was run by Tony Coleman, an entrepreneur slash community organizer with a nonprofit called One Fam. The organization got its start organizing against police violence in the '90s. One Fam also fixed bikes for kids and maintained this coffee shop. Tony favored track pants. People around here knew him as a good guy.

The cafe, however, had almost nothing for sale. One time I went in and a Katrina refugee in a black beanie named Cedric was tending the place. He was serious, smart, dark about America as befitting someone who got out of New Orleans because he thought the levee was going to break. He liked to call other Black men "Black man," as in "How you doing, Black man?"

I told him that I was working on this book and he started talking about technology immediately, how it felt so crazy, this world that has been built. "Technology always gonna have a downside," he said. He compared it to sci-fi shows. "It's like when they go out in space and shit and they be looking for stuff and then they bring it back here and the people say, 'We didn't want that shit! Why you go out looking for that shit?'"

No one asked for this world and yet here we are. A BART train screeched by, trucks rumbled. I was the only customer.

I wanted to support the place, so I tried to order food, but although there was a whole menu on the wall, all they actually had was coffee and day-old coffee cake. So I bought those.

In this local world, business worked like this. You bought something for a few dollars— coffee cake, coffee—and then you sold it, maybe, for a few dollars more than you bought it for. Do it enough times and you can pay Cedric. Do it a few more times and Tony Coleman can keep the doors open. This is the economy as we like to imagine it.

But this is not how everyone makes money. My first job after college was with a hedge fund. In my first week, one of the founders had lunch with all the new people. He said something like, "I've never wanted to make anything in my life. I've just wanted to use money to get more money. If that's what you want to do, this is the right place for you." It was not the right place for me, and I lasted just a few months before quitting one glorious morning in 2004.

Places like private equity firms, hedge funds, and VCs take in money from a set of financial institutions that manage the real big money: public and private pension funds, Middle Eastern sovereign wealth funds, insurance companies, huge asset managers, university endowments, banks. Some of these places have half a trillion or even a trillion dollars under their control. BlackRock had $9 TRILLION under management in 2023. Even the small ones might have $25 billion to play with.

Combine it all and that is the "giant pool of money," as *This American Life* famously described it during the financial crisis of 2008. These dollars represent capital that can be invested just about anywhere. The program's reporters pegged it at $72 trillion back then. In 2023, the world's largest asset managers have more than $113 trillion under management.

Then, as now, this capital was willing to scour the earth for a return on these vast sums. As American banking deregulated, large banks became a conduit for this money to flow into anything that could be "securitized"—i.e., bought and sold like a commodity.

Outside the giant and global pool of money, the technology firms draw on global markets, too. While places like the Revolution Cafe have to rely on individual residents of West Oakland to come in and spend three dollars so they can make fifty cents, a company such as Alphabet or Meta can show ads to the entire world. They extract money from the internet and pull the bulk of it to their headquarters.

Or take Uber, which took in billions of dollars of investment in the 2010s, a large chunk of it through the Japanese SoftBank, which itself drew funds from Saudi Arabia's sovereign wealth fund. Uber was losing money to keep fares low to juice its growth; in a very real sense, when you took an Uber ride, some percentage of the cost of your fare was subsidized by pumping oil out of the ground in Ghawar province of eastern Saudi Arabia. How can a regular cab company compete with that?

Even the companies operating out of the Port of Oakland have almost no relationship to the neighborhood economy. The shipping and stevedoring outfits are headquartered in Asia or Europe. The goods they import and export almost all come from somewhere else. Oakland captures some tiny trickle of funds, but the big money flowing through the port exists almost on another plane of economic existence, inaccessible to anyone in West Oakland except through the very narrow conduit of increasingly automated dockwork, where the longshoremen capture maybe 1 percent of the value of the goods passing through as wages.

Battles over urban change often take on a cultural cast—coffee shops and dog walkers replacing BBQ joints and barbers. The real disagreement, though, is not over lattes but whether the development model predicated on driving up property values works for everyone. It's whether a rising tide lifts all boats. In city after city after city, the answer is the same: the rising tide is a flood and it does what floods do. Some people drown. Others run. It is more than metaphor that the homeless encampments, filled with flotsam and jetsam, look just like the remnants of a natural disaster. This, too, is an emergency.

If you're poor, there's nothing you can do about it, except wait and watch for the signs of change, disturbances in the wind or the taste of the air that say, "Get out now." The name that people have given to this terrible fluttering unease about the very centers of American capitalism is gentrification.

20

The Base

Phil Tagami leveraged his downtown development success into a deal with the city that fulfilled one of his longest-held dreams: rehabilitating the old, grand Fox Theater. Again, the development money largely came from grants, tax credits, and a $13 million loan from the city. Again, Tagami succeeded where others had failed. Despite the subprime crisis and economic chaos, the Fox reopened in 2009 as the true cornerstone of downtown Oakland's resurgence.

In twenty years, he'd come up from nothing to perhaps the most important developer in the city, someone who had mastered how to get things done in a notoriously fractious city.

Uncle Phil was the toast of the town. Rumors were floated that he'd run for mayor or city council. And he did confess to me that he once thought he would run for mayor. He held court at Oakland nightclubs, dined with local politicians and celebrities, and was not above visiting the glorious dive bar Cafe Van Kleef.

Tagami had made it, a crucial causal force in Oakland, as well as a major beneficiary of the incipient revival. Downtown Oakland still had empty buildings, but it had been set on a new trajectory that would eventually transform its skyline. No single person could claim more credit for that than Phil Tagami, one way or the other.

Tagami is not a traditional fat cat. He was born in Oakland and grew up mostly in the east, an area called Dimond Heights. He played lacrosse and got a scholarship to a boarding school in Maryland, before returning to Oakland and dropping out of high school. Even his friends don't know the whole story. He is an autodidact (one who, when I used that label, said, "Good word. Greek") and someone who is keenly aware that he is operating in a world that people with his background normally don't move in. Once, when we were talking about urban development philosophies, he offered up the quip "If you're gonna brush with Robert Moses, you have to rinse with Jane Jacobs."

His father was Japanese, by way of Hawaii. "Frank" Tadao Tagami was a mechanic who worked at a variety of shops around the Bay Area, memories that have stayed with Phil.

His mother is white and her people were from down Arkansas way, in the south-central part of the state, just an hour or two from where Margaret's mother had lived. They, too, had been migrants, and they married at a time when miscegenation laws were just coming off the books. The couple were not always welcome in the white parts of town.

College, even Margaret's partial, fractured, multi-institution version of it, was not in the cards for Phil Tagami. He worked as a roadie, setting up drums, and in construction doing labor before he met his business partners working at the athletics club.

Tagami loved rugby and yoga—even co-owns a studio in Oakland—and in recent years, he wore his thick hair long with a goatee and round, dark-rimmed glasses. He tweeted Bukowski poems ("a poem is a city filled with streets and sewers filled with saints, heroes, beggars, madmen, filled with banality and booze") and business koans ("breaking things takes no talent, building things that last does"). A reporter I know who covered him extensively remembered him as "crazy, but interesting."

Let's be real: most real estate developers are not exactly wildmen. They're business guys. In the Bay Area, most of them are white. A huge part of the game is accessing large amounts of capital, and being a plausible recipient of vast sums of other people's money may be the most important white male privilege. But in the Oakland that created Phil Tagami, being a stuffed suit who could funnel money into construction projects was not enough. It took a bit more courage, even recklessness, to try to build in a city that was perpetually falling apart.

Given his countercultural leanings, Tagami was a perfect counterpart to Jerry Brown. Brown, who'd been governor of California in the 1980s and had run for president three times, lived for a time in a warehouse in Jack London Square and hosted a radio show before he decided to run for mayor of Oakland. Tagami was a crucial part of the campaign as a "behind the scenes powerbroker" cultivating support. Brown won the mayoral election and trampolined back to state politics, winning election as attorney general and then getting a second go at the governorship.

As Brown came to power, the biggest, most intractable project in Oakland remained the Oakland Army Base. It'd been fifteen years since the closure decision was made. And more than ten since the community plan failed. Scheme after scheme popped up. Perhaps it would be a row of car dealerships or a luxury hotel or even a mall. The wildest plan came from the Wayans brothers of *In Living Color* fame. They would put a large film studio out there and employ the people of Oakland making movies. But nothing panned out. The land sat and sat and sat.

Until Phil got involved.

His company, CCIG, rejected the fancy commercial development idea that had come into vogue, offering an alternative that was referred to as the "working waterfront." "We are pursuing a new economy," he said in one news article, "but also one that respects the need for a living wage and one [workers] can support a family on."

The port deal was particularly complicated because although the city of Oakland got all the Army base land, the management structure for the project had become diffuse. The land had been split, so the project would run in one city, but through two different bureaucracies.

Perhaps that made sense at some point. If the port was doing maritime stuff and the city side was a movie studio, why would they need each other? But if the city portion of the Army base focused on logistics, too, then the whole area would need to be master-planned with common infrastructural improvements. Someone would have to bring everyone together, track down grant funds, bring in private money, and negotiate with the thousand community, environmental, and labor groups likely to want a seat at the table.

Really, there was only one man for that kind of job, and it was Phil

Tagami. He'd been a port commissioner, and he had friends at every level of government, all the way up through Jerry Brown in the governor's mansion and Barbara Lee in Congress. Oakland Global, their operating brand, adopted the slogan: "One Vision, One Project, One Team."

The linchpin of the project was a $242 million grant from the California Transportation Commission using dollars known as Trade Corridor Improvement Funds. That didn't just happen. Tagami went out to town after town all over Northern California—Ripon, Fresno, Galt, and others—building support for Oakland to get those dollars. Tagami's team would write letters supporting their priorities in exchange for their backing for the port project. "An unprecedented coalition of Bay Area and San Joaquin Valley transportation agencies declared the project the number one priority for all of Northern California," the executive director of the Transportation Commission at the time recalled.

The dollars were slated for massive infrastructure improvements that the port and the city would share. Then, in phase two, the two entities would develop on their own, but working toward building a synergistic logistics cluster. Tagami had even managed to land Prologis as a partner in the new waterfront hub, which should sound more impressive than it might to you. Prologis is the kind of company that's everywhere, a crucial component of the global economy, and yet it is nowhere. If you walk down around the buzzy Ferry Building in San Francisco, its logo sits at the bottom of the Port of San Francisco signs, an unexplained but clearly critical part of the infrastructure. Its main offices are in the first pier to the north of the Ferry Building. From the unassuming building, Prologis has come to control a network of 1.2 billion square feet of warehouses. It focuses on warehouse space close to cities, which is one reason it is Amazon's largest real-estate provider, and Amazon its largest customer.

In 2011, Amazon had 48 million square feet of real estate. Not a small number, but nothing like what it would need to cover the earth with its own warehouses. Over the next seven years, Amazon would add 240 million square feet of real estate to its holdings. It needed companies like Prologis to build out or buy warehouses and get them ready for e-commerce use. In 2018, Prologis bought another warehouse company for $8.5 billion. In 2019, it bought yet another for $12.9 billion. In 2020, another for $9.7 billion. And another for $22 billion in 2022. The company likes to say that 2.8 percent of *global* GDP runs through its warehouses. In the United States, the value of things passing through a Prologis warehouse is equal to 36 percent of all the goods consumed in the country. That's how crucial a piece of the Pacific Circuit it is.

And even back in the early 2010s, it was a grade A partner. If Tagami could get the port to sign off on a plan to do the city and port development together, he'd have the kind of project that would vault him to global status.

But there were some problems, and a big one was Ms. Margaret Gordon, who didn't trust Phil Tagami, nor did she think more growth at the port was necessarily a good thing.

←---   ---→

As Margaret's first term as a port commissioner came toward its end in 2011, she and her allies imagined she'd be reappointed.

But things got messy.

The Teamsters' hopes for organizing port truck drivers were dashed by a loss at the Supreme Court, but they swung their organizing momentum into work on the Oakland Army Base. They created a community coalition with EBASE and others called Revive Oakland to push for more local, union hiring on the massive construction project that the Army Base promised to be. People were talking about 1,500 jobs, at a time when Oakland had not really recovered from the Great Recession.

Margaret had her own community coalition around the Army Base called Oakland Works that wanted a lot of the same things that Revive Oakland did. Both wanted massive numbers of local hires and a "ban the box" provision to prevent formerly incarcerated people from getting their applications tossed in the trash. Margaret and the West Oakland Environmental Indicators Project wanted more emphasis put on the environmental concessions. They'd spent way too much time cleaning up pollution problems in West Oakland to allow new ones to be created through the Army Base project. The WOEIP and Alameda County proposed a wildly inclusive stakeholder process for monitoring emissions for the project. But they got slapped down and shushed.

So, again, Margaret found herself not exactly opposing the Teamsters but certainly in a different camp.

Oakland's mayor was Jean Quan. She became the first Chinese American mayor of a major city, a couple years before Ed Lee became the mayor of San Francisco. Quan had been involved in the Berkeley formation of the Third World Liberation Front, then worked in New York's Chinatown. Many imagined she'd govern as a fairly radical mayor in the post–subprime meltdown era.

But even her early supporters in Oakland began to question her politics. In an essay in the Asian American magazine *Hyphen*, Chris Fan, one of the publication's founders, saw her as "signal[ing] the incoherence of 'Asian American' as a radical coalition." In the wake of her treatment of Occupy Oakland protesters, Fan said, "No other public figure dramatizes more powerfully just how distant those heady days of action and idealism have become."

When it came to the Army Base, Mayor Quan wanted people on the Port Commission who would play ball with the city's big plans for the base, which meant playing ball with Phil Tagami leading the development of the joint infrastructural phase of the project.

"Why did he have to have the whole project and why did the port have to raise money for him to do the development?" Margaret asked. One reason: Jerry Brown was in the governor's mansion. Brown had long been a Tagami ally, and that $242 million of state money was supposed to arrive for the project.

The pressure to move Margaret off the board started to build. The Teamsters were only too happy to provide some names of acceptable possible replacements. "I was asking too many of the right questions," Margaret said.

At first, Margaret successfully fought off being booted from the board. Mayor Quan appointed a young man, Jakada Imani, to replace her in the fall of 2011. "Was that dirty?" Margaret shouted at me, remembering it. "That was dirt-ayyy!"

Imani was an impressive young man. A protege of Van Jones, he'd worked for Nancy Nadel in West Oakland. He was a young up-and-comer with a politics that seemed more in tune with the rest of the Oakland labor alliance.

Margaret's supporters were having none of it. The longshoremen came out for her. One even said, "In the ILWU, we consider her a part of our family." Nancy Nadel made an impassioned speech in Margaret's defense. "I know how important it is to the West Oakland community to have that voice on the Port Commission, finally, after all these years. And we shouldn't let it go," she said. "I can't remember any other port commissioner serving only one term. And that's outrageous to me." She listed Margaret's accomplishments, noted how she'd become a nationally recognized expert on the ins and outs of logistics, and pulled in major national prizes. Nadel called her a "bridge builder" who "really brought the truckers and environmentalists together for the first time ever."

Mayor Quan said she hadn't meant to pit progressives against each other. Her selection was meant to bring balance and youth, while sharpening the board's sense of itself as a team. Labor's annoyance with Margaret had nothing to do with it. The port had major competitive problems. It needed a board that would help Oakland "capture the investment from China." She depicted Imani as a compromise between a more militant figure like Margaret and having "just businesspeople."

Imani himself had to admit before the city council: "Margaret Gordon is a giant and there is no denying it." To cheers, Nadel chided him, saying she was "a little disappointed that he's letting himself be used in this situation." After a weird council snafu in which he was confirmed and then that confirmation was invalidated by the city attorney on procedural grounds, he never came up for another vote because he pulled his name from consideration.

Margaret was safe for a few more months.

But then the Teamsters found their man: Cestra Butner, a Maserati-driving business owner who'd created a major beverage distribution company in Oakland. Butner might have been a representative of Black capitalism, but he ran a union shop, organized by the Teamsters, and they liked the guy. They brought him to the Alameda County Labor Council and won an enthusiastic endorsement from the rest of the unions. At a July 2012 city council meeting, union rep after union rep came forward to pledge their support for Butner.

Josie Camacho of the Labor Council and Greg McConnell, a lobbyist who works with big businesses in Oakland, stepped to the podium together. "We are in full strong support of Mayor Jean Quan's appointments," Camacho said. "When do we have a situation where we have business and labor literally standing next to each other?"

On July 25, a *San Francisco Chronicle* column made the stakes clear. "There are also rumblings that the mayor's Port Commission picks would give her the votes she needs to help developer and City Hall insider Phil Tagami win the full development rights to the old Oakland Army Base," two columnists wrote, "a deal the current Port Commission has blocked."

It was a signal. The fix was in, despite all the nice things the councilmembers had said about Margaret just a few months before.

Her last stand came at the city council meeting on July 26, 2012. Only four speakers came out against Margaret: Doug Bloch of the Teamsters, Camacho of the Alameda County Labor Council, McConnell, and a lawyer for the Oakland Chamber of Commerce. All were respectful of Margaret's tenure. They could afford to be big. They'd won. Everyone knew what was happening. Big labor and big business were combining to throw the one community advocate off the board. The redevelopment project was complicated enough without a powerful West Oakland resident pushing for more accountability. Bloch put it in stark terms: they needed new port commissioners to ensure that hundreds of millions of state infrastructure funding would get put to use. "I don't need to remind you that $242 million in state infrastructure bond money is at stake and not taking action on these appointments puts everything at risk," he said.

Her people would speak, though.

Brian Beveridge, her longtime partner at WOEIP, approached the podium in a collared shirt and mussed hair. He had written something for the occasion. "Removing Margaret Gordon from the Port Com-

mission silences not only the voices of West Oakland residents bur-
dened by the pollution, traffic, and noise of the Port, but it silences the
voice of the environment and public health," he said. "For more than
seventy years, the Port of Oakland has been directed by the voice of
business, and for most of that time by the interests of the conservative
white system of economic power in this country." On the video of the
meeting, Beveridge looks heartbroken, but not surprised. "There is a
fragile truce between the Port and the people of West Oakland," he
concluded. "Removing Margaret Gordon from the commission will
break that truce, but then again, breaking truces with the poor and disempowered is one of the things our government has always done best."

Then came Monsa Nitoto, part of Margaret's cohort of environmental justice leaders. He was dressed immaculately in what can only
be called an all-white zoot suit. His tie matched his pocket square; his
hat was on. A thick man, Nitoto quavered with emotion. He had been
in the trenches for so long. And he wanted to talk about that history
a little, specifically Mayor Jean Quan's role in the different political
machinations they'd engaged in. Nitoto talked about how Quan had
"knocked down the Black population in Oakland," and how they didn't
have the power they once did. "Margaret is the representative," he said,
and then began to cry. He stepped back from the podium. He'd given
up so much for his activism. The man had many health problems and
his businesses had never really worked. What he had was his legacy,
which Margaret, in part, represented. He buttoned the top button of
his suit, recomposed himself, and stepped back to the mic. "You going
backwards," he said.

Person after person got up attesting to Margaret's great qualities.
No one was really opposed to Ces Butner as a human. He was well-
liked, just like Jakada Imani. But they all affirmed that Margaret was
an independent voice who stood for the West Oakland community, no
matter what.

Margaret did have one union in her corner. ILWU Local 10 sent its
president, Richard Mead, to speak for her. He approached the microphone in one of the union's windbreakers, two pens tucked into the
pocket of his white shirt. He told the story of a warehouse they were

trying to organize. After weeks, they got a hold of a pay stub and found that the guys were already paying union dues, but the union had done so little, they had no idea. "This is what happens when management and union get in bed together. I'm not saying that's what's happening in this case, but it sure has that appearance," he said. "If Mr. Butner was a member of the Teamsters it'd be a totally different story. He's on the other side of the table. And one of the things we've learned in the ILWU is: *stay on your side of the table.*"

Margaret spoke last. She came out of the audience with her close-cropped hair, gone thinner over the years. Normally stylish and matchy, she'd dressed conservatively for the occasion: a dark pantsuit and a frilly pink shirt.

"I been here before. This is my third time," she said, "and I've never told my story." She spoke in measured tones, a final statement more than a plea. "All I wanted to do was do service," she said. "I've served my community and I think I've served my community well." The commissioners had all either voted for her in the past or said they'd support her. And here they were, about to replace her.

"The story of me becoming a commissioner has been fascinating to many people throughout this country. They want to know how a *little Black woman* from the neighborhood or ghetto of West Oakland got in this position. And it was about hard work," she said. "I didn't know anything about air quality. I didn't know about research, about data, about community participatory research."

From 1989 to 2012, Margaret had reinvented herself, dedicating herself to her West Oakland neighbors, to environmental justice, and to figuring out how to deal with the repercussions of a necessary but damaging part of the global economy. In those twenty-three years, her body had begun to go on her. She was barely able to walk because of cleaning so many houses, stooping, spraying, mopping, dusting. And yet she was a giant, even her opponents had to admit. You couldn't get around her, because she was not just angry but savvy. She could see the machinations of a political leader like Jean Quan as if she were x-raying the city. She knew how power worked. How could she not when it had been exercised against her so many times?

"Emissions is down fifty percent in West Oakland," she said. "That didn't just happen!" This was her moment to take credit for what she'd contributed over the years. The countless, countless meetings at dinnertime. The reports she had to read. The meetings with young white girls from the environmental movement who had no idea what she was about. The people who thought any brown person talking about green was a "token" spokesperson for the real researchers and thinkers doing the work. All that. But Margaret also knew she had not done it alone. She hadn't even done it first. Her victories were communal. "That happened because there was a whole group of us working together over the last twelve years," she said.

She turned to Mayor Quan. "I voted for you. I got my son to vote for you. I voted for you because your son came to my front door," she said. Hadn't Margaret sent her own son out into the streets of West Oakland to stand on street corners, counting trucks? She and Quan were both outsiders, really, to these power structures, even if they now found themselves inside them.

"Mother to mother, I have to appreciate our sons," she said. "I just hope that everybody else will do the same thing for me, recognizing all the hard work I have participated in and tried to do for West Oakland and communities like West Oakland throughout this country. Thank you very much."

Fighting back the pain of movement, she walked away from the podium, shifting her weight from one leg to the other. Most of the crowd rose, person by person, to applaud her. There were women like her in so many Black communities, but at the same time, there was only one Ms. Margaret of Oakland, California.

It was the city council's time to vote. They were visibly ashamed. Desley Brooks, the only Black woman on the council and the only one who'd voted against confirming her years before, delivered a scathing indictment of her fellow council members. "I remind you, each and every one of you, about your word," she said, pausing. "If I had been thinking I would have asked KTOP [the city TV station] to cue up that meeting, where I watched each one of you talk about how 'if Margaret's name were brought forward, I would reappoint her. I'd have no problem doing that.' Each and every one of you." The audience heard the call and began to respond. One man cried out, "Tell 'em sister."

"So what does your word mean?" she asked, the crowd grunting a loud assent. "What does your word mean? Who are you?"

Council president Larry Reid had the answer. The council members were politicians bent to the power assembled around them, a unified front of big business and big labor, who wanted Margaret gone. That nearly unprecedented alliance is what it took. "I understand the political consequences of my vote, whether it is labor not endorsing me, or the Oakland Metropolitan Chamber not endorsing me," Reid said. It was that kind of vote, the one the powerful would remember, an are-you-with-me-or-against-me litmus test. Almost no one wanted to fail it.

"Margaret, you know how much I love you and how much I respect you and the work you've done," Reid said, "not only as a port commissioner, but one who cares about others." He ultimately abstained.

Love wouldn't save her. The tiny giant had fallen.

21

Big Shit

With Margaret out and new appointees on the board, Tagami's path was clear. All the various factions came to a kind of detente—including the WOEIP, for what else could it do—signing on to the Project Co-operation Agreement, a massive document laying out the community benefits from the redevelopment. "We were able to win fifty percent local hire on that project," the Teamsters' Bloch said. "We were able to win restrictions on temp agencies. We won a job-training center. We won apprenticeships requirements, project labor agreements. The whole thing is being built union." These were important things. The lease agreement alone was 1,700 pages long.

For EBASE, the Teamsters, and Jean Quan, it was a legitimate victory. Margaret and her crew had meaningfully contributed to it, too, but their reward never did arrive.

The Army Base project ceremonially broke ground on November 1, 2013. The power structure behind the redevelopment emerged to take a bow for bringing nearly $250 million in spending attached to a bunch of solid local job provisions. A giddy Fred Blackwell, who had negotiated the deal for the city of Oakland with Tagami and CCIG, introduced Phil Tagami. "I cannot think of a better person to lock his jaws on a project like this," he said, before noting that everyone already knew who Tagami was, and canning the rest of his intro.

Tagami wore a simple gray suit with a bright white shirt, open at the collar, cuff-linked. "It's been said that the goal was to make sure that everyone got what they wanted," he began, "and that's ultimately not a realistic goal." In balancing the real demands of the project—commercial, environmental, community—there had to be compromise. "There is going to be dissent. There is going to be an active, creative process. That is community action. It's not community apathy, it's not community torpor," he said. "It's engaging and participating in the process, respecting other opinions and voices, and coming together and finding that common ground."

In fact, none of the politicians on the stage could quite believe that they'd gotten the project going. Tagami had a colorful metaphor for describing its difficulty. I'd told him that I imagined he felt like a guy in a strongman competition pulling a semitruck. But no, he said.

"It was: we want you to shit in a swinging jug, okay? And we're not going to tell you the speed of the jug. All we're going to tell you is you need to eat all this food, right? You need to take all the stool softener in the world. We're going to put a blindfold on and even spin you in a chair. Then there is going to be a gauntlet of people with bats with nails in them that are going to hit you to slow you down. And then when we say shit, you shit in the swinging jug and you need to get all the poop in the jug," Tagami said. "We did that."

The newly elected city councilwoman from West Oakland, Lynette Gibson-McElhaney, noted, "This has languished for fifteen years." And given the difficulty of the project, pushing it ahead could be "called a total victory. And it is OK to celebrate that today."

Representative Barbara Lee reminded the hometown fans that she was bringing home the bacon for her district in the form of a federal grant. And Jerry Brown got up to crack some jokes about how old the base was, which is to say roughly as old as the governor himself. "All this nonpartisan stuff is nice, but at the end of the day, it's your friends that take care of you, and you've got a friend in Sacramento," Brown concluded his brief remarks. People cheered wildly.

Margaret did not trust Phil Tagami. They'd signed on to the umbrella agreement to help push things along, but word of a new Tagami plan began to spread.

There are different kinds of port facilities. Oakland primarily had *container* facilities, but Richmond, for example, was largely an oil port, where tankers fed crude oil into huge tanks. There are also "bulk" terminals that deal in things like grain, ore, petroleum coke, animal feed, and, of course, *coal*. There are several bulk terminals up and down the West Coast, but none in Oakland, even though it is a main export node to Asia because its "drainage" includes much of interior California.

Part of the dizzying array of plans for the new Port land included an entitlement to build the "Oakland Bulk and Oversize Terminal" (OBOT) out on the city's side of the old Army Base. There were few details available about it, but environmental activists smelled a rat. A bulk terminal? Anyone could look at a couple of charts and see that coal was a huge part of the bulk commodities market.

As the Army base redevelopment went ahead, the Port of Oakland was exploring what to do with the Howard Terminal, an old part of the docks that it wanted to redevelop. When it put out a call for ideas, it got few responses. The only serious proposals came from Bowie, the coal company, and Tagami's CCIG with a group including Kinder Morgan, a massive fossil fuel transportation company. That sparked speculation that, if the port turned him down, perhaps Tagami would take the plan over to the Oakland Army Base, where, after all, he did have that entitlement, and which just so happened to be the kind of place you could ship coal from. Sierra Club investigators filed public records requests with the Port of Oakland, which produced documents that led them to believe that CCIG had coal plans at the Army base.

Looking back through the documentation around the Army Base project, it is pretty clear that the city knew or should have known that coal was a possible commodity at a bulk terminal. But Pete Cashman, who was managing the Army Base project for the city, said that it was obvious coal would be a politically difficult commodity in a legal deposition. And that Tagami seemed to agree it would be "stupid."

In December 2013, Tagami's Oakland Global newsletter tried to tamp down the rumors. "One bulk material OBOT does not plan to export or import is coal. CCIG and Port of Oakland officials have been asked about potential coal shipments as part of Oakland Global and

OBOT. Coal is not in the plans, according to Tagami," the newsletter stated. "'It has come to my attention that there are community concerns about a purported plan to develop a coal plant or coal distribution facility as part of the Oakland Global project," Tagami said. "This is simply untrue. The individuals spreading this notion are misinformed. CCIG is publicly on record as having no interest or involvement in the pursuit of coal-related operations at the former Oakland Army Base."

In January, Tagami posted on his Facebook page that OBOT was "saying NO to coal as an export product. We are committed to emission reductions here and abroad. We share this one planet and the only path to clean the air is to at some point stop polluting it." In early February, he tweeted a chart depicting energy source statistics. Coal and other fossil fuels were still going up and to the right. He noted, "We must accelerate the use of modern renewables!"

Later that month, the port rejected all proposals for Howard Terminal, specifically because of the coal that it might have handled. While Bowie was upfront about it, the port noted that CCIG had not included any information on which commodities it might handle, but "staff infers that commodities similar to those proposed by Bowie Resource Partners may be handled under this proposal."

The issue fell off the radar. Tagami seemed to have settled it: he wasn't doing coal. And that did kind of make sense. After all, his most powerful political ally, Jerry Brown, was making fighting global warming a key part of his platform as governor. As far back as the 1980s, Tagami had been exposed to the ideas of environmental justice and, like most Oakland residents, believed that climate change was real.

At the same time, there were factors in the Pacific Circuit that made coal a more attractive export option. Namely, the earthquake and tsunami that led to the meltdown at the Fukushima nuclear plant had led the Japanese to pivot hard away from atomic power and toward a next generation of fossil fuel plants. They were suddenly hungry for coal, and were racing to build dozens of new power stations. That didn't go unnoticed by Tagami or his team.

← - - -   - - - →

Brian Beveridge and Margaret may seem like an odd duo to have created the West Oakland Environmental Indicators Project in its current form. She's extroverted, passionate, and a Bay Area native. Brian is soft-spoken, analytical, and from elsewhere. She's Black, he's white. It works.

They both believe in data as a tool in the struggle. They commit to long, long, long processes and to understanding every minute detail of the systems they are trying to change. They build relationships across racial and class lines, bringing in rich Black people and poor Latinos, white labor lawyers and Indian environmental activists, tech bros, teenagers, and seniors. "We have individuals who think that we pulled a rabbit out of our ass. This was work. It was built on relationships and trust," she told me. "I will work with anybody. I may know you a asshole but I will work with you, if there's an opportunity to get you and others on the right path."

There's a famous story about Ms. Margaret. Gathering their own data is a key part of WOEIP's method. Air quality is a very local problem; pollution can vary widely, even within blocks. But the way governments measure air quality is to stick a single sensor in a neighborhood and figure it does a good-enough job of capturing the data.

People at the Environmental Defense Fund had dreamed up a beautiful new way of getting better data. They'd mount pollution sensors on Google Street View cars already bristling with cameras, and then send them driving around, taking high-resolution readings across Oakland, including Margaret's corner of the city.

Data-loving Google got on board. This fit squarely into its mission of organizing the world's information. A company called Acclima built the sensors. By spring 2015, they were ready to roll.

Ms. Margaret warned the outsiders that they should do outreach, let everyone know what was going on. The street would notice, she told them. But Google's and EDF's people knew in their hearts they were doing something good for the community, and they pressed on. The plan was to run the cars on most weekdays for roughly a year. They'd build a massive dataset of hyperlocal air quality, unmatched in its spatial and temporal resolution.

In the spring, the cars began to roll out of garages in Mountain View and San Francisco and make the drive over to Oakland, where they'd spend six to eight hours zipping around.

After a while, though, some residents began to get suspicious. In particular, the drug dealers working at the then-bustling corner at Eighth and Campbell. Day after day, a car bristling with cameras comes rolling through? Something was up.

So, eventually, the guys stopped a young woman who was driving one of the cars and asked her precisely what the fuck she was doing there with all those cameras. She got scared and Google pulled its drivers off the road.

Of course, everybody called Ms. Margaret. She gathered the Google and EDF people, stuck them in a van with her old friends Richard and Gloria, and drove the two minutes over to the trouble spot.

The men on the corner were her neighbors, people who'd walked her home from house calls back when she worked in the local school. They might have asthma, or their kids did, like so many of the young people in Prescott.

So, when Ms. Margaret limped up to the men holding down the corner, she was received with kindness. She explained that the people there with her were the ones running the cars they'd seen. "They is not taking pictures of what the fuck y'all doing out here. My name is connected to this," she told them. "You know my name is not gonna be connected to anybody's BS." They hugged Ms. Margaret, thanked her for her service to their kids and community. The Google and EDF staff hung back, eyes bugging.

Then she went around the way to the man selling perfumes and explained it to him. Then to the liquor store and the child-care center and the Campbell Village projects. There's no surveillance attached to this, she said. It's about air quality.

The Google cars had no more problems in the streets after that.

Ms. Margaret knows everybody.

←----   ----→

For all the public speaking she has had to do, I don't think Ms. Margaret likes having a mic in her hand. She'll do it, but she says what she came to say and gets off the stage. In our many interviews, I could grasp how she thought about Oakland, but it was always a piece of another story, another planning process, another bit of history. She never led with her mission statement or ideology. What I learned was that, in their relationship, Brian often plays the Aaron to her Moses, speaking aloud their shared analysis. He'll often open meetings laying out the big picture, while Margaret works the room.

Environmental justice groups are famous for *blocking things*. That's one of their core functions. But as Mari Rose Taruc, a Richmond EJ activist, put it, the strategy has to be: "Fight the bad and build the new." I understood their fighting, but what was their philosophy of building? What was their alternative to the status quo? Brian laid it out for me one time, as we sat in the shabby conference room of the trailer where they work. He was wearing a light purple faded long-sleeve Sundance T-shirt tucked into his jeans, a crazy belt buckle, and a beat-up brown leather jacket. First, we talked about the gang war in the '90s between the Acorn projects to the east and a smaller tower closer to BART. Eventually, the city razed the smaller project and scattered the families across East Oakland. Then it built back a smaller number of affordable housing units and brought in mostly better-off people.

As we dug deeper into the history of the area, we started talking about a series of anthropological and archaeological studies that Caltrans had engaged in after the fall of the Cypress Structure. They'd been useful to me in understanding the history of West Oakland, I said, but it didn't really seem to help the community to have created all this knowledge.

"I think the tragedy of that is that you have there the seeds of actual development—being able to do development from the ground up, which is, in my opinion, the only way to moderate gentrification. It's kind of like the story about the gangs in the housing project. Rather than go to the people with these problems and figure out what their needs are and provide for those needs so that they can become functional, we tear down the place they live so they have to move away and we can build something new and we can sort of control who moves in. And we think because we got rid of the worst-case offenders now we've solved everybody else's problem by giving them subsidized housing. If you really wanted to do a needs assessment and an analysis of how to build people up, you would want to understand what makes them tick. You would want to do an anthropological and cultural analysis. How do you function? How does your community work? What is your ethnic cultural foundation? Who do you trust and who do you not? And how do you talk to each other? What do you own and why? Where does your financial support come from? Besides, you know, we give you a check, so we pretty much understand, right? No, you don't understand. You don't know how when the check doesn't last all month, like it never does—how people don't go broke and starve to death. You don't know the informal economy that takes place in places like West Oakland. So not knowing any of that, our solution to poverty is: let's bring one hundred million dollars from somewhere like China and build stuff. Because if you invest, obviously, that makes the world a better place."

Brian was speaking softly in long, complete paragraphs. Outside, Ms. Margaret was listening to smooth jazz, which filtered into the room.

"But there is no method for the people who live here to ride that wave. All boats don't rise, because the people here don't have a boat. They just stand there while the water rises around their necks," he said, anger rising in his voice.

He paused. I hadn't said a word.

"So I think, you know, it's a long way of saying, I think all this research should have helped, but for the most part, it was done in some kind of academic way. Let's check off a box we were required to. We have to do our anthropological analysis of this place before we bury it under asphalt. Literally. You know, because there's probably history down there and people that expect us not to just pave over their history. We have to hire some people from the university. But it doesn't mean anything. No one's ever gonna go through it as a source for anything except historic reference. They could have said: Gosh, who lived here? Who lives here now? What brought about the transition? What do we understand about economics and life and joy and peace based on this history? And how could we use public funds we're going to use anyway to enhance the economics and the peace and the joy and the quality of life of people who live here? I think with all the complaining about growth and gentrification, we actually don't do anything about it because we just do the same stuff over and over. We just think, 'Where's another forty million dollars?' Let's build another thing. You know, the city wants to take, what, fourteen million dollars and buy a defunct hotel and build it into homeless housing or something. I mean, the city hasn't spent fourteen million dollars in West Oakland in twenty years. As Margaret is fond of saying, we spend two hundred fifty million dollars on infrastructure at the Army Base, sewer lines and sidewalks and pavement. They haven't spent a dime on that stuff in West Oakland, where the people live. But over there for the trucks and the warehouses, we're going to gold-plate the public infrastructure and then expect that to somehow pay off for the people of West Oakland. I mean, I guarantee you could give every unemployed person in West Oakland one hundred thousand dollars. It would cost a little less than what they spent on the Army Base and all their lives would be better."

I told him that one of my central questions in the book was why money could flow *through* West Oakland in the form of cargo from the port or government dollars or even the drug trade, but it didn't seem to stick in the area. The obvious structural answer was racism, but it felt like it was hard to go deeper than that and say, well, then who *is* making the money that could go to help people in West Oakland?

"That's why Tagami is such an interesting character," I said. "You can see this is a person whose goal is to tap the public coffers and pull the money out of there—"

"To make things better," he said. "I don't fault Phil. Phil has become a millionaire doing it, but I don't think he doesn't believe that he's trying to make things better. They all do that. Who was the guy who rebuilt New York City? Robert Moses. They all think they're making it better. But it's all trickle-down theory. They wouldn't call it that. But it's all like: if we build big shit, somehow that serves the needs of the poor. When the poor have nothing to do with big shit! They're not even in the system. That's where the racism parts come in. Saying it's all about racism is true, but it's kind of like being out in the woods surrounded by lumber. And the sun's going down and you say, 'Man, it's getting cold. Why is that?' I mean, even your friend says, 'Because the sun's going down.' It's like, 'Oh, we're cold.' Well, why are we so cold? Because the sun went down. But no, we're actually so cold because we're not building a fire. We're not addressing the issues. We're blaming a systemic problem without saying: How do you be sustainable in this country? How do you build wealth and sustainability in our economy? You own stuff. In the classic sense, you own the means of production, which you're going to own in a business which produces capital. You don't have to produce a wheelchair or, you know, automobiles. If you own the way to produce capital, you will have equity, economic equity. Or you own real estate. If you don't own either of those things then you can't produce capital or leverage wealth, leverage value by owning something you could borrow against, you're nowhere and you're going nowhere. Because jobs come and go in today's world. They want us to be perpetual employees, perpetual renters in this society, so that the people who do own stuff will control all the money. So if you look at West Oakland, and say, well, how do you fix it? People say we've got to fix racism. *Well*, that's a heavy lift.

"But if you said the result of racism is that nobody got to buy any houses, we could work on that. The result of racism is that there are so few locally owned businesses that are owned by people of color in this neighborhood that serve the people of this neighborhood. And so churn those dollars over and over, keeping it in the neighborhood and hiring the kids from high school as stock boys and hiring the local people as hairdressers. If you did that, you would be offsetting that racist history that denied people access to the mainstream economy. But we don't do any of that. We will help Phil Tagami to the tune of two hundred fifty million dollars, but we think it's a big deal to lend the local woman who wants to open a hair salon fifty thousand bucks so that she can open up a brick-and-mortar space on Seventh Street."

22

Keep It in the Ground

Utah has tremendous coal reserves, but as environmentalists have pushed the electricity industry away from burning rocks, the mining industry in the state has been in serious decline. The problem isn't that the miners don't have the resources but that they can't find a large enough domestic market to make mining coal profitable. The fossil fuel is, more or less, stranded underground, which is very good news for climate change. There is no realistic way to fight global warming without stranding *a lot* of coal assets. Or as Ms. Margaret's five-word position about coal goes: "Keep it in the ground."

Richfield, Utah, is perhaps best known for being approximately halfway between Los Angeles and Denver, but the small town is also located in a valley on the western edge of Utah coal country.

For people in rural Utah, a thousand miles from Oakland's coast and culture, coal is a local industry worth promoting. So perhaps the editors of *The Richfield Reaper* did not think much of a story they posted on April 7, 2015: "Project Could Transform Local Coal Market to International." It described a plan for four Utah counties to put up $53 million for the express purpose of gaining access to an export terminal in . . . *Oakland*. Bowie Energy had purchased a mine called Sufco in Salina Canyon, and wanted to ship its coal globally. "The proposed port would be located on a bow tie shaped piece of land in San Francisco Bay. Formerly occupied by an Army base, the land is owned by the city of Oakland," the newspaper noted. "The company, Terminal Logistics Solutions, has signed a 66-year lease to develop the property into a port."

Small though the story was, tiny though the town remains, the paper's publication set off a series of explosions out in California. For just about everyone in Oakland, this was the first thing they were hearing about Terminal Logistics Solutions. And here was $53 million coming from a government agency for the express purpose of sending coal through the Army Base.

In spring 2014, Tagami's company had entered into an exclusive negotiating agreement with a new entity staffed by two former executive directors of the Port of Oakland, Jerry Bridges and Omar Benjamin, the latter of whom was bounced after a scandal about visits to gentlemen's clubs. Nobody had noticed a new California business registration in the Rotunda building, Terminal Logistics Solutions, LLC.

After the *Richfield Reaper* article hit, the Sierra Club hit Utah with a series of public records requests, one of which turned up a smoking gun from Jeffrey Holt, chairman of the Utah Transportation Commission. "We've had an unfortunate article appear on the terminal project . . . If anything needs to be said, the script was to downplay coal and discuss bulk products and a bulk terminal," Holt wrote in an email. "The terminal operator is TLS, not Bowie. Bowie is known for coal . . . Phil Tagami had been pleased at the low profile that was bumping along to date on the terminal and it looked for a few days like it would just roll into production with no serious discussion." (Tagami, for his part, said he never said anything of the sort to Holt.)

If I can summarize the feeling among people who'd worked with Tagami over the many years of the Army Base project, it would be simple: *he fucked us.* Sure, he *could* have exported coal, but no one thought Tagami would be crazy enough to actually do it! Oakland is, after all, probably the most liberal city in America. Containers fine, coal no.

As the bad press spread, Bridges tried to finesse the situation. He headed out into the West Oakland community, offering to pay a few cents for every ton of coal shipping through the area into a fund that local groups and churches could use. They even solicited Margaret and Brian at a lunch at Nellie's Soul Food. Margaret had a long history with Bridges. When he was working as the port's executive director, she'd cornered him after a meeting. Bridges is a very big man, maybe six feet two and built strong. "I went up to him and I bump-bump-bumped my chest into him. And he looked down," Margaret said, laughing hysterically. "I said, 'My name is Ms. Margaret Gordon.' The man said, 'I have never had no little woman, no woman run up in my chest like that outside my mama.' When I told him who I was and what I was doing, all he could say was 'Yes, ma'am.'"

Which is to say, Bridges probably should have known better than to dangle money in front of Margaret. They turned him down right away and got in touch with Darwin BondGraham, then a crack reporter for the *East Bay Express*, who covered the story closely. He was able to confirm that Bridges had been making the rounds of local churches, talking about money and jobs for the community. Some pastors, in a neighborhood with nearly 30 percent Black unemployment, even got behind the terminal.

The city as a whole, though, turned against the project. State legislators started talking about bills to complicate things. The city council swung into action. Even the former mayor Jean Quan, who had worked hand in glove with Tagami, said, "The approval process would have been very, very different if Phil Tagami would have said, 'We're going to do coal.'"

Despite the public opposition, it was immediately clear that the situation represented a very big problem for the city of Oakland and its activists. Those deeply involved knew that Tagami had a strong hand on the basic legal issue. "The city council, by incompetence or by naivete or whatever reason, really gave the store away to Tagami in their contract with him," one anti-coal activist told me. His rights out on the Army Base had no restrictions on the commodities he could ship, except, as Tagami put it, "nuclear waste, illegal immigrants, weapons, and drugs." If Tagami and Bridges and Bowie wanted to do coal, they could do coal.

Because of that, the case against Tagami had to turn on the one exemption in the development agreement, often referred to by its number in the document: 3.4.2. It said that the city had to protect the health and safety of its residents. That one provision was their only hope. So activists and the city council alike coalesced around the idea that if they could show that there were health and safety impacts to the people of West Oakland, then they could stop the bulk terminal. After all, in 2014, the city council had passed an ordinance banning the transportation of "hazardous fossil fuel materials" through the city.

But given that the city had allowed West Oakland to be polluted for decades and given that the coal trains *might* add only mildly to the burden of the neighborhood, they needed to put together evidence that this particular project would hurt residents. Hundreds of people testified to the city council. Thousands of pages of written documents were submitted. Most focused on the deleterious effects of coal dust coming off the trains. The project's backers countered that the coal cars would be covered. Anti-coal activists pointed out that not a single coal car in the whole country had ever been covered.

Against the advice of the No Coal in Oakland activists, the city commissioned a hastily composed report by a consultant called ESA. Other groups submitted evidence about coal dust's repercussions, but the ESA report would eventually take the starring role.

As the coal export controversy heated up, Native American activists in the center of the country went after another fossil fuel transport project: the Dakota Access Pipeline. A bundle of new technologies, fracking most prominently, had opened up vast new oil and gas fields in the United States. But, like the Utah coal miners, the developers needed new infrastructure to get their product to global markets efficiently.

Environmentalists realized, like the longshoremen before them, that the transport of fossil fuels created choke points where they could exert power to frustrate the powerful interests. There is no scenario in which the world burns all its coal reserves and does not derange the climate to an extreme extent. For climate activists, then, coal exports are a cardinal sin. In terms of the climate, it doesn't matter where on earth greenhouse gases are produced. As Tagami had once put it: "We share this one planet."

Coal usage peaked in 2007 in American power plants, and was in the midst of a free fall that's seen its use to produce electricity drop by more than 60 percent. Natural gas, made cheaper by fracking, and scaling wind and solar industries ate up market share. Coal companies were going bankrupt; coal mines were shutting down. Without exports, their mines might have close to no value. This was Bowie's incentive to fight. "The rough justice of the market reaction is a red flag," one energy analyst wrote after a Bowie deal collapsed. "No public money should be invested in mines that may be worthless and in a company with such a troubled past."

But, if it could find new markets in Asia, like Japan, it might be sitting on a half-billion-dollar set of assets.

Opponents of the project have said that it alone would increase U.S. coal exports by 10 million tons, or what they say is 20 percent. In a global context, the numbers are large, but not staggering. The United States consumed 500 million tons of coal per year, and globally, humans burned more than 8 billion tons. And not to put too fine a point on it, but it is coal that charges up the grids of Asia, so its factories can manufacture the goods that come back in containers through the Pacific Circuit. We export our emissions to Asian manufacturing economies. All that to say: the stakes were not small. There is a reason that up and down the West Coast, environmentalists had fought new bulk terminals for exporting coal with everything they had.

With these national and global dynamics swirling around them, at a packed meeting in June 2016, the Oakland City Council unanimously voted to ban coal handling in the city, a measure obviously targeting the Army Base project. This was a big victory for local activists, but they knew that it was not over. After several months of failed negotiations, Tagami filed a lawsuit.

"We applied for and the City approved and vested exactly what the market demands: a terminal capable of being fully responsive to market demands for global transport of legal commodities over its 66-year useful life," Tagami said. He had a point: market logic had governed everything else the city had been doing for decades. Why should this be different?

Suddenly, Margaret had a lot more friends, who now needed her help organizing to put pressure on Tagami. A coalition of environmental organizations, from the national Sierra Club to WOEIP, with a group of anti–fossil fuel activists from Richmond called the Sunflower Alliance, came together calling themselves, simply, No Coal in Oakland. As the face of West Oakland environmentalism, Margaret grounded the broader coalition in the community and prevented Bridges from framing the issue as white outsiders preventing the creation of jobs in the Black community. You couldn't out–West Oakland Ms. Margaret Gordon. Plus, history had shown that Margaret had been right not to trust the Army Base project and its financing.

The dispute would have its day in court.

In January 2018, I rode my bike down Mandela Parkway, toward a No Coal in Oakland meeting at Margaret's office. They were getting ready for a protest outside a hearing at the federal courthouse in San Francisco, where Judge Vince Chhabria would or would not deliver a summary judgment on the case. It was going to be their big day in court, and I was excited to hear about the planning.

But as I got closer to Seventh Street, I saw an enormous police presence. Maybe twenty police cars were blocking Seventh and Eighth. Legions of cops. Regular people milled about, too, a sour hum in the air. The police were blocking the little parking lot across from the BART station. "They won't let me get my car and I need to go pick up my kid at school," one woman said helplessly. News helicopters began to buzz overhead. I asked a photojournalist slung with cameras hustling by what was up. He confirmed what I'd guessed: there had been, in the dry words of bureaucracy, an officer-involved shooting.

A BART police officer had killed Sahleem Tindle, a member of a large and connected Black family in the Fillmore. The scene had been simmering for hours. Exactly nine years and three days before, BART police had killed Oscar Grant at the Fruitvale BART station, touching off months of protest.

Television reporters approached Truck, the owner of the Upper-Kutz barbershop near the BART station, and asked him what he saw. Truck, however, wanted to tell them what he'd seen at a deeper level, all of which he broadcast on a Facebook live stream, as he stood in a black hoodie in the rain.

"I don't appreciate how the news and how the media portray my people," he said. "And I don't want to give my words for a news segment that's gonna get chopped up and twisted around." Two men had been in a struggle when the BART police officer ran up and shot Tindle right in front of Truck's shop. He'd seen it all, but he refused to turn his experience into content for local news, who'd done enough damage. "I don't think the news is going to help Black men be OK. It ain't never helped us," Truck said. "I ain't never done nothing violent and when I finish watching the news, I feel like a damn criminal."

The police portrayed Tindle, who was walking with his partner and children, as the aggressor in the conflict. It was not even five blocks from where Huey Newton got into his gun battle with Oakland police.

For months afterward, Tindle's family, usually dressed in all white, would lead protests around the BART station, at BART board meetings, and in other government buildings. Cat Brooks and her organization, the Anti Police-Terror Project, would shut down BART board meetings and lead marches. As usual, the policeman in question, Joseph Mateu, was never disciplined. As usual, BART—which is to say everyone who rides BART trains—would eventually have to pay out a huge settlement. Cat Brooks would run for mayor, launching her campaign from Tony Coleman's place on Seventh Street, and garnering a solid 25 percent of the vote.

Even before I pieced the whole story together, there was a lot swirling in the air that night. Anger, history, confusion, and, just down the road, Margaret and her colleagues hard at work inside the trailer.

I arrived at the No Coal in Oakland meeting late. Everyone was gathered around a plain conference table. Mostly, it was older white people. A few other organizers joined via the phone. Margaret sat at one end of the table in a white T-shirt, a furry white bucket hat with a bow perched on her head. Tacked up on the wall was Tagami's quote about CCIG's lack of interest in a coal project at the Oakland Army Base. Two young people sat across from Ms. Margaret, a blue-haired twentysomething named Carlos Zambrano and a high school climate activist named Mykela.

Ted Franklin, a former labor lawyer and self-identified ecosocialist, spoke. He had longish gray hair and a small mustache. He laid out what was going to happen. Everyone was hopeful that the judge might rule in their favor. They were downright optimistic that the whole coal project, which they had been fighting for a long time, even then, might be over.

The worst-case scenario would be that the judge would open up the evidentiary record and relitigate the city council's determination that the facility might pose a risk to residents.

But, if they did not win that day, the organizers had other plans. "Even if Tagami wins, even if he wins at the Supreme Court four years from now, coal is not gonna go through Oakland," Franklin said. "We have other strategies."

Aside from organizing local protests and keeping up pressure on the city to keep fighting in court and otherwise, Franklin and the rest of the team had developed a strategy for following the money. Every source of dollars and every waypoint along the path was a potential target for investigation and pressure.

Everyone knew the early money on the table was supposed to come from Utah, but No Coal in Oakland researchers had dug into the details. Jeffrey Holt, an investment banker with the Bank of Montreal's investment banking wing, BMO Capital Markets, was the central figure in the deal. He'd consulted with the four Utah counties and put the deal together while he was chair of Utah's Transportation Commission. At the same time, he promised to place $200 million in bonds for the project in his role at BMO. Was this a conflict of interest? The activists thought so.

The $53 million from Utah, though, came from a public entity called the Community Impact Board, which gathered fees from mining on federal lands to distribute to local Utah governments. Supposedly, these funds were to *mitigate* the effects of extracting fossil fuels. Was funding an export terminal in California a form of mitigation? No Coal in Oakland activists did not think so.

The Bank of Montreal, for its part, had signed on to the Equator Principles, a voluntary set of risk management guidelines for financial institutions. Just as an example, signees were required "to demonstrate effective Stakeholder Engagement as an ongoing process in a structured and culturally appropriate manner with Affected Communities." Had the Bank of Montreal come anywhere close to meeting the requirement?

Holt had put out an info sheet on the "multi-commodity bulk terminal" for prospective investors. It said the project had received "broad Governmental support," and predicted it would throw off tens of millions of dollars each year, without once mentioning the word *coal*. The big buyers of this kind of bond are pension funds, part of the giant pool of money. And many of them are public, like the California Public Employees' Retirement System, which has more than $350 billion to throw around *and* a series of environmental and sustainability goals. Would pension funds investing in Holt's bonds *know* that they were supporting a coal export terminal, and if they did, how many people would want to touch it, given the local and national opposition to the project?

Franklin argued that banks were likely to try to raise an "infrastructure fund" with pension money, then simply roll the export terminal into the mix. It all smelled a bit like the mortgage-backed securities: take something risky and mix it with other things and the known risk dissolves away.

The activists discussed protest plans, arguing over music and turnout. Margaret wasn't convinced that an 8:30 a.m. rally in San Francisco was likely to draw as many people as they wanted. So they should plan on a small, tight rally, rather than overextend. She would speak at the rally for a few minutes and handle press interviews at the courthouse.

They concluded the meeting precisely two hours after it began, as planned. They scooped rice and beans Margaret had made onto paper plates. The No Coal folks went out into a night still buzzing with helicopters.

On the day of the hearing, No Coal activists unfurled large signs that read, KEEP IT IN THE GROUND and TAGAMI v OAKLAND—THREAT TO HEALTH, CLIMATE, DEMOCRACY outside the imposing San Francisco federal courthouse. Supporters straggled in, each identifiable with a red "No Coal in Oakland" T-shirt. It was not a huge rally. I counted about sixty participants. Most were boomers, aside from a contingent of students. Some had been fighting Tagami on the project for almost five years. They had helped dig up the health-and-safety loophole and worked to make sure evidence of a "substantial" threat to the public made it into the record. They thought, that bright morning, that perhaps a real victory might be at hand.

Margaret wore a long black hoodie and a red cap. Interviewed by a local TV news reporter, she delivered her messages. She noted the general pollution of West Oakland, the health impacts to children, and the climate angle. "Coal needs to stay in the ground," she said. "It needs to stay in the ground. We have no real use for it. We have renewable energy processes and technology and coal needs to stay in the ground."

When it was Margaret's turn to speak, she started with a simple call and response. "I'm about keeping it in the ground," she said. "Keep it in the ground!" And the crowd chanted back, "Keep it in the ground!" West Oakland was already overburdened, and this coal terminal sure as hell wasn't going to help. "The Army Base was supposed to be a benefit to the community of West Oakland and we are not benefiting from the storage of coal. We have health issues. We have people and children who are sick and have asthma," she said. "We do not need coal to be stored in West Oakland. Keep it in the ground!"

The protest wrapped up and we ambled through the metal detectors. The courtroom itself was cavernous. No Coal in Oakland supporters packed the back benches. A few besuited supporters of the terminal sat close to the front.

From the beginning, Judge Vince Chhabria ruled the courtroom imperiously. Almost immediately, it was clear that Chhabria *was* going to relitigate the health-and-safety evidence that had been presented to the city council. Chhabria's approach seemed to be that everyone else was kind of an idiot, so he would have to figure out, for himself, if the coal project posed a "substantial" danger to West Oakland residents.

In a bruising back-and-forth with the city's outside attorney, Kevin Siegel, Chhabria pushed and pushed on the *relative* harm of the coal export terminal in such a degraded area. Siegel, at first, responded that there was no safe level of small particulate matter, PM2.5, and therefore the city need not worry too much about the details of how much pollution might escape the trains heading to the terminal or the facility itself. But Chhabria cut him off.

"If there is a justification for banning this activity, it cannot be the thing you repeat over and over in your brief, and the experts repeat over and over in their reports, that there is no known safe level of exposure to PM2.5," Chhabria chided. "It cannot be that."

The community was already heavily impacted by pollution, Siegel sputtered back. And the city council, at their discretion, could make the decision to protect residents. Chhabria cut him off again to force the discussion about the Oakland Bulk and Oversize Terminal's importance in the city's air quality picture.

"What is the effect of other things that happen in Oakland compared to what would happen at OBOT? How much PM2.5 pollutes the air from the Bay Bridge toll plaza? If a new stadium were ever built for the Oakland A's, how much PM2.5 would pollute the air from the presumably three-year stadium construction project?" Chhabria asked. "How much PM2.5 pollutes the air from 880 going through Oakland? What about the demolition of a building and the reconstruction of a new high-rise building? How much PM2.5 pollutes the air from that? One of the things that seems to be missing from the materials that have been submitted and that were considered by the Oakland City Council is *how big of a deal is this* in terms of the amount of PM2.5 that's going into the air and dispersing in the air compared to any number of other things that happen on a daily basis in Oakland?"

Siegel had no answer. Tripping over his words, he tried to argue that the rest of the city's air quality should not preclude the city from taking action in this case. The judge would not relent. He presented

Siegel with a hypothetical. "If every other activity that takes place in Oakland results in the emission of ten million tons of PM2.5 into the air and the activities at OBOT result in ten tons of PM2.5, that additional ten tons," he said, "does that additional 0.001 percent create a substantial danger?"

No Coal in Oakland activists were shifting uncomfortably. This was one of the core arguments of environmental justice completely inverted. EJ advocates often note that the cumulative impacts of lots of different pollution sources make addressing each of them more important. That was one of the conclusions of Tony Iton's research in the area. In many cases, it wasn't one pollution source or another that killed people early. It was all of them together, that weathering.

Judge Chhabria was saying the opposite: if there is already all this pollution, then what's one more source? He tried to force the city to admit there was some pollution threshold under which OBOT could operate that would not pose a danger to the citizens of the city.

"How do we understand the magnitude of the danger if we're not comparing it to all the other sources of pollution?" he asked. Murmurs periodically went up through the courtroom.

There *could* have been such a measure, of course, but the city had not wanted to use one. Because the truth was: the city had let so much pollution rain down on West Oakland already, it *was* legitimately hard to argue that *any* source of PM2.5 would be outlawed. And the city was *actively* involved in growing the Port of Oakland. The rest of the infrastructure that Tagami's team had built at the port would almost certainly generate more emissions in the neighborhood.

A Sierra Club lawyer tried to step in to stem the bleeding. She noted that there were national air quality standards—and that all the reports submitted to the city found that OBOT's operations would make it more likely for Oakland to go over those thresholds on any given day. Chhabria countered again that the warnings were vague. She made one last go, explaining that other courts had found enforcement action was justified at the rough magnitude of pollution that all sides in the trial agreed the terminal would generate. Chhabria was not satisfied, and he sent everyone to lunch, knowing that there would not be a summary judgment, or if there was, it would not go well for the city.

Outside the courtroom, Margaret was spitting mad. "This is the bullshit with the city coming back to bite them in the ass," she said. Activists had wanted it to bake air quality standards into development deals, but it had not wanted to do so. "They have never established a standard on a per-project basis for air quality standards working with public health data," she said. And given that the city had not banned other developments, Margaret saw that Chhabria had a point.

The journalist Darwin BondGraham walked by. "They really shot themselves in the foot there," he said. Margaret, still pissed, shot back, "All the way up to their asshole." Which made perfect sense, even if I still haven't figured out the geometry.

The judge ruled that the two parties would have to go to trial. Throughout the ensuing days, Tagami dressed in beautiful, perfectly tailored suits, pinching the bridge of his nose in the universal sign for being tired and stressed. In the lunchroom one day, Tagami arrived at the same moment as a pack of T-shirted No Coal in Oakland activists. First, he bought a No Coal T-shirt from one of them. Then, as they brought wilting institutional salads to the registers, he ostentatiously ordered a double bacon cheeseburger.

His own testimony was unremarkable. Tagami's lawyers did a good job of painting many of the city's actions for what they were: a late-in-the-game attempt to create the evidence to block a massively unpopular development. Without climate change as a key driver for fighting the coal terminal, the trial devolved into a battle over the minutiae in hastily written reports, and to no one's surprise, they didn't hold up very well.

When the trial ended, it seemed clear that the city had lost, but Chhabria's May 2018 opinion was vicious. He called the record before the city council "riddled with inaccuracies, major evidentiary gaps, erroneous assumptions, and faulty analyses, to the point that no reliable conclusion about health or safety dangers could be drawn from it." Outside observers were aghast. "You hardly ever see an opinion that is so one-sided," a local law professor marveled. "The judge goes in chapter and verse explaining what was wrong, and you could hardly make more mistakes, it looks like, than what was done here."

For all the effort that local activists had put into shutting down the development, they could not overcome the city's incompetence in putting together its ordinance correctly or in developing a case that could pass the evidentiary bar. On the face of it, it was not clear that OBOT would be even a small player in the air pollution picture of West Oakland. The legal fight wasn't over, but the cast of it was established.

In essence, Chhabria's decision reaffirmed the status quo: the city had been willing to sacrifice this neighborhood for economic reasons for decades. Why stop now?

After two and a half years of my trying to interview Phil Tagami, he agreed to talk with me. We set up a time in November 2019. Phil was, more or less, Margaret's nemesis, but he was also an Oakland booster. I had no doubt that he wanted the best for the city, even if he pursued different ends than the community activists or local politicians. I was excited to get his version of events.

Then, on the day we were supposed to meet, I got a call from his assistant. "Phil's in the hospital," she said. "He's fine—but he's in the hospital." I offered to reschedule, but she told me, "No, he specifically told me to keep the meeting."

So, on a sun-drenched late fall day, I set out on my bike for the Summit Hospital, just off Broadway, near where thousands of new condos had recently replaced a row of car dealerships and repair shops. The hospital lobby was bathed with golden light through its large windows. Room 4763, I was told, and took the elevator up. Right at four, I knocked.

"Come in," I heard.

Inside, there was Phil, in a baggy T-shirt and shorts, sitting in a wheelchair. His left arm was in a sling.

With some relief, I figured that he'd gotten into a bicycle crash or maybe broken his collarbone playing rugby or something. I looked down at his bulging calf muscles, then out the window, at a jaw-dropping view of the city, the port, and on to San Francisco and the Golden Gate. The light was settling onto the buildings, golden mist.

I turned back to Phil and asked him what happened.

"I had a stroke," he barked.

I did not know what to say.

It had happened the week before. He'd woken up in the middle of the night unable to feel his left arm. He thought it was just asleep, but then there were no pins and needles. Always the autodidact, grabbing his phone with his good arm, he googled symptoms of a stroke and called his doctor, who sent him to the emergency room. At first, his left leg was working at about 70 percent, but it got worse over the days, and by the time I saw him, it was down to 30 percent. I caught myself before asking if they thought he'd make a full recovery. He thought he'd be out in a week, perhaps, and in rehab for seven months.

The sight of him hit me harder than I expected. Phil was robust. Phil was a vital force. A heart attack, sure. A car accident, definitely. But this, it seemed like the poisoning of a gladiator before the battle. It was not fair. The man was a lion and here he was pushing and pulling his left arm around like a dead fish.

If he was afraid or shaken by the experience, he didn't show it. As I had him recount the past, he was enthusiastic. He had an old friend with him in the room, and his business partner, Mark McClure, who eyed me warily as I sat down. Sipping sparkling water, two Tupperware containers of kale chips in front of him, he told me about being a drum-mer and a roadie for Santana and Joe Satriani. We walked through his early career, and the '89 earthquake.

About twelve minutes in, he made the turn in the conversation. "Let me answer some questions for you," he said, his voice cracking a bit. "I'm getting a little tired."

That's when I started to get Tagami's side of the story on perhaps the most controversial project in Oakland history, and certainly of Phil's career. It had turned him from the consummate insider and one of the city's *true characters* into a pariah.

From Tagami's perspective, it was clear what had happened. He'd signed a contract with the city that allowed him or his lessees to export whatever they wanted. Everyone loved the idea of having a "working waterfront," one where people without fancy educations could find good jobs as the labor force attached to the Pacific Circuit. The revenues from that terminal were a crucial part of making the whole Army Base deal work. And they'd somehow managed to drag that whole damn thing almost to completion when everyone else had failed.

And then some bad press came out about coal possibly getting shipped through the commodity terminal and that scared everybody. For Tagami, the real problem was that the Oakland city politicians were afraid of the Sierra Club. None of the city staffers who'd worked on the project could admit that they'd known coal was at least a *possibility* to be shipped through the bulk terminal when they'd negotiated the deal.

Tagami had timelines and maps spread out over his hospital bed, and he began showing me binders and charts with exhibits that he would later email to me. Tagami had many beefs with the city, but most important, he felt like his dignity had been impugned. The city, he felt, had thrown him under the bus. It had known all along that he was planning a commodities bulk terminal. Half the bulk commodities market is coal. Every comparable port shipped coal. Their first possible terminal operator was Kinder Morgan, which specializes in the transport of fossil fuels. The word *coal* was lurking in its communications with the city as far back as 2011. Sure, it would be more profitable to ship other things, which have a higher price per ton, so its terminal operators would try to ship those things. But the bulk commodities market wanted coal, too, so they'd end up shipping some coal.

The city might not like that they were dealing in coal. But a deal is a deal is a deal. Legally, contractually. If the city didn't want them to ship coal, it should have put that in the original contract. Or, if the city later decided that they shouldn't ship coal, it could and should have renegotiated the lease with OBOT to reflect the reduction in value of the terminal because of the restriction. Instead, the city passed the anti-coal ordinance, then decided to try to defend it legally. At this point, it was just delaying and delaying, a kind of financial siege warfare.

"They literally told me, 'Just go away. Just forget about it,'" he told me. "I'm like, 'Excuse me?' After ten years of my life and thirty million dollars spent and I'm supposed to just go, 'Yeah, oh, forget it.' Ten years of my life!"

It occurred to me, as he buckled his left arm into the sling, and then unbuckled it and tossed it onto the large armrest of his wheelchair, that perhaps those ten years suddenly seemed more precious.

I asked him about his declaration that coal would not be shipped through the terminal, made years before. Tagami called it "true when spoken," by which he meant that at the time, they had no prospective terminal customer who was interested in shipping coal. Was that misleading? Perhaps. Was it a lie? Tagami says no.

More interesting to me is how Tagami felt about the moral side of shipping coal. He was not a Trump supporter. "I believe in climate change," he told me. But did that mean that at *that moment in time* when Japan was shutting down its nuclear facilities and building coal plants, *he* should not be able to lease his terminal to someone who would send coal, and cleaner-burning coal than Japan could otherwise get? Jet fuel and gasoline generate carbon dioxide emissions, too. Should you not be allowed to fly to Mumbai or drive across the country to Florida? The Sierra Club contends that the scale of the project makes these comparisons absurd.

Tagami's answer is basically what the head of Japan's largest coal power plant operator told the *Japan Times*: "We share the same principles as others to realize a low-carbon society. But we think low carbon is a little different from a coal-free society." Japan's new coal power plants are substantially cleaner for the local environment than previous generations, too, although their carbon dioxide emissions far exceed natural gas, to say nothing of nuclear, solar, wind, and other non-fossil technologies.

Was the world coal-free? Was Germany still burning and exporting some of the dirtiest coal in the world, despite passing the most progressive renewable energy policy?

He was passionate, but he'd also been through all this before, clearly. Activists had protested at his house. They'd protested a fundraiser for his kids' school, which ended up being canceled. Someone who had been known, respected, and even feared by all the powerful constituencies in the town had become, in his own word, "radioactive." This was a man who had come up with nothing in his city, and working off the plans that the city made had actually delivered for Oakland and himself. How could he not regret having taken on this project?

"I'm happy where I'm at, including me being in this chair right now," he shot back. The sun was setting behind San Francisco. Our beautiful town was bathed in that orange we get in the late fall, early winter, the kind that lights up the windows on the houses in the hills.

That his greatest political ally, Jerry Brown, was pushing renewable energy was beside the point. This would be a profitable endeavor, and the city could not simply outlaw a certain economic activity it didn't like. "Restricting any commodity on political grounds puts a cloud of uncertainty over the entire project going forward," Tagami said in a legal filing.

The market had governed everything else the city had been doing for decades. All the redevelopment, the urban renewal, the port. Wasn't it all trying to stimulate economic growth? Even the whole Army Base deal itself. Hadn't Margaret's environmental concerns been run over by the promise of "good paying" jobs? Did the local government believe in allowing the market to shape the city or not? Tagami was following the logic of the times.

PART IV

Circuit City

Noni Session

23

In the Interest of Others

In the months after that first Margaret gumbo party, as the sides of the coal dispute sparred, I returned to West Oakland again and again. I'd sit on a concrete bench outside the post office and just watch what was happening around me. I'd ride my bike around the streets. I'd interview people who lived nearby. I'd dig into the Instagram and Facebook posts of residents. I plotted the homes going up for sale and the developers who were planning bigger things. I'd ask about certain buildings, certain streets, the old train station, the BART, the post office, Mandela Parkway, the railroad, the trucks, the different public housing projects. And then I'd go find the receipts in archives and libraries, tucked into oral histories or long-forgotten planning documents and PhD dissertations.

I had already learned the things that everyone learns about Seventh Street, the lingering memories of the blues clubs and the complete Black society that grew up in West Oakland. I read a couple dozen books by and about the Black Panthers. I got drunk with the old longshoremen and kept up with some of the tech people who were scattered through the area. I went to meetings and events put on by the phalanx of nonprofits operating in the city. I tracked down Phil Tagami's history and tried to get him to talk to me about why the hell he was trying to export coal through Oakland.

It all felt like a big story about capitalism and what a city is for, and what the powerful owe to the people who make a place alive. One day, I was walking past Slim Jenkins Court, when an exterior door on Seventh popped open. A person's possessions came flying one by one onto the street. A distraught young woman was leaving her boyfriend. I held the door open as she brought the rest of her things onto the sidewalk. Her auntie somewhere up I-80 called her an Uber. We loaded everything into the trunk as she fought back angry tears. "I can't wait to get the fuck out of Oakland," she said, slamming the door on her way out of town.

Her *Oakland*, the way it sounded in her throat, was a different place from the one I knew, the one most people I knew knew. We lived in beautiful Oakland, in one of the hottest housing markets in America, in one of the world's most diverse cities, in a progressive bubble, in beautiful housing stock, in blooming gardens, in fresh coats of paint, in remodels, in "everyone is moving to Oakland" Oakland, in "we got pushed out of San Francisco, too" Oakland, in "we're trying not to be that kind of gentrifier" Oakland.

The city was changing fast in those days. Crime was down, unemployment was down, property values were up. Libby Schaaf was the new mayor in town. Schaaf read as a soccer mom, but she'd been close with longtime Oakland politicos. She'd been an aide for Jerry Brown and chief of staff for Ignacio de la Fuente. Then she ran public affairs for the Port of Oakland before launching her campaign. Schaaf won a messy election that saw the longtime Tagami ally Bryan Parker's campaign stall out and Jean Quan struggle. The buzzword around the mayor was *techequity*, which I'm not sure anyone ever pronounced it "te-kwitty" as Mayor Schaaf did.

The concept did get at a crucial dynamic in the city, though. For the first time since 1950, more white people were moving in than out. The young tech workers who had built Facebook and Google and the iPhone were aging into parenthood, and Oakland's bungalows and Craftsman housing stock, built for city escapees not unlike themselves during the 1910s and 1920s, became a hot commodity. Bushrod, where Bobby Seale's parents lived, was on its way to becoming the single hottest neighborhood real estate market in the country. Uber had even purchased an old Sears store in the heart of uptown and became the first major tech company to land in Oakland . . . ever.

As in previous eras, Black people were the early adopters of the future economy, and it *sucked*. The Pacific Circuit was driving the mutation of cities—the trade in containers and the currency exchanges and the demographic shifts and even the coal. The Pacific Circuit was running a cargo ship through the former reasons for cities to exist: manufacturing things, housing the people who worked there, and providing services for them. This slice of the West Coast seemed to be reaching a breaking point. What are cities becoming when the big money is made inside a screen or manufactured somewhere in Asia? What is a city for? And what does Oakland owe to the people who live here, rich or poor?

Rumors and plans for new buildings were everywhere—condos out along Broadway and wrapping down to the lake. Huge new developments down by the water south of the port. And towers. Real big towers for a downtown skyline that had been frozen for years. What role was there for either a "working waterfront" exporting coal or a Black resident of public housing like Ms. Margaret in a city like that?

I remember my first Uber ride. It was probably 2012, and I was running late for an event in Mission Bay, the new master-planned area south of the South of Market neighborhood. I'd planned to walk, but I was running out of time, and San Francisco (notoriously) was short on cabs. I'd heard about Uber and, as my train pressed through the underground, downloaded the app, hit the button, and by the time I ascended the escalator into the drugged-out human misery of the Sixteenth Street BART stop, a car was pulling up. I got in. The driver was friendly, or perhaps courteous, more properly. Five minutes later, I arrived, early. There was no awkward exchange over the tip or protestations that the driver did not take cards. I just got out. It almost felt like a video game where a nonplayer character, an NPC, just appears to do your bidding, helping you along in a quest. Which should have felt dystopian from that very first time. But it did not.

The whole exchange was the most perfectly convenient thing. Hit button, get in car, arrive at destination. So simple! So clean it hardly felt as if money had changed hands. The driver, sitting on one side of this marketplace, had gotten paid whatever Uber said he would. The passenger, me, sitting on the other side, had paid whatever Uber said I would.

Uber, like Facebook and Amazon and Google, attained ubiquity so quickly that the radicalism of the model hadn't had time to really sink in before the organization of urban transportation had been transformed. This was a new kind of company, jet-fueled by venture capital and representing, quite plainly, a future that Milton Friedman only dreamed of. This was a company that had devoured supply and demand, brought the market in-house, and used it to supply labor without employees, rides without cars. The app was the boss. The company was a thin technical and managerial layer, like a private equity firm with a teensy-tiny bit invested in a million drivers. For a user, it was a personal logistics system: you delivered anywhere, or anything delivered to you.

Uber—the idea as much as the company—is an argument about how the world should be organized, who should benefit, and what connections people have to one another. Uber doesn't really hire or fire people, nor does it tell them where or when to work. It simply tunes the market, laying the physics that sends cars to different spots, that pulls in people from deep in Contra Costa County to the metropole.

From a corporate perspective, this was unprecedented, though in the years since Uber launched, hundreds of companies have risen with a similar model. As Uber drivers began to coalesce as a group, mostly online in forums but also at coffee shops and airport parking lots, they began to go: hey, our earnings have gone down since the early 2010s.

The one time I got to interview Uber's latter-day CEO, Dara Khosrowshahi, I asked him about increasing drivers' take-home pay. His answer was astonishing, when I really thought about it: "In general, if rates overall go up, demand goes down, and if demand goes down, driver utilization goes down, and then overall earnings often go down or don't go up," he said. "There is actually very little that we can do in terms of overall earnings for drivers."

It was a bit like the commissioner of the NBA complaining about the rules of basketball. Uber paid its engineers hundreds of thousands of dollars, paid hundreds of millions of dollars to acquire a small number of engineers working on self-driving cars, expanded wildly across the world, lost billions of dollars of venture capitalists' money, and also tinkered constantly with the set of financial incentives that would get drivers out on the roads. It financed many of the cars drivers used. It decided on the classes of service and managed the reputation system. Uber itself controlled the market conditions with the opacity of the golden-age Kremlin.

Yet the CEO of the company actually said that there was little he could do to change the earnings of the people doing the work of his company, whether he was willing to call them employees or not. What more perfect example of liquid modernity could there be? Anything can happen and nothing can be done.

I'm not interested in Uber per se. Its failings have been detailed elsewhere in all their gore and grimness. But the worldview that Uber represents is extremely common. It says: Solidarity is an illusion. Community doesn't matter. Everything is transactional. Anything can happen, but nothing can be done, at least to improve people's lives. How did this universalizing Uber individualism come to be?

There is an intellectual history there running from Milton Friedman and his University of Chicago colleagues to Stanford's Hoover Institution libertarians and into Silicon Valley. Cabdriver regulations were one of Friedman's favorite specific punching bags. Friedman's libertarianism remains a powerful solvent that can be poured over any attempt at rising above narrow self-interest.

Solidarity is hard. Organizing people is hard. Being committed to other people year after year is hard. Hitting a button on your phone is so easy, and it increasingly became the way that wealthier people in cities did everything.

One day, years before the pandemic, I drove east through a gap in the East Bay hills to see a new industrial warehouse site that Prologis was building. The project manager toured me through this new logistical terrain where, quite recently, cows had grazed. We came upon a massive new concrete pad. The workers around its edges looked like they were on the shore of an unnaturally smooth gray lake. Walls would be tilted up soon, and it would become another small node in the Pacific Circuit.

We drove on to a recently completed warehouse, empty except for light and dust. It stretched off into the distance. The guy managing the project looked at the vast blankness. "People don't realize this is where the future is," he said. "No one's going to shopping malls. Shopping malls are going into here, right?" Amazon, not an entity known for subtlety, has begun to buy up dozens of malls around the country to use as distribution centers.

Out at the old Army base, there would be bright white lines of Prologis warehouses, too. The logistics cluster would come together, bulk terminal or no. So, in a sense, the mall that Oakland planners once imagined might go out at the Army base has arrived.

But for a city, there's a huge difference between a retail shop and a warehouse. The former is a place, a spot on the map. In the best of cases, a community can develop around it. A warehouse, located in a port industrial district or inland off some highway, offers none of those positive externalities. It does not add to the life of *the city*. People don't come together in these spaces; they are merely delivered to from them.

That's a fundamentally different way of organizing the commerce of a city, and it is worse for the fabric of urban life.

In the morning, the archetypical tech worker queued up for the bus and traveled down to Facebook's arc, which faces the great mudflats of the bay and which most days is a disconcerting brown. He need not go outside, though; he could travel for miles *inside* the office. The design suggests the starship *Enterprise*'s fractal endlessness, but also the old ladies who colonized the mall, when that was a retail necessity, to get their daily steps.

Or maybe he arrived at Apple, part of the company's elite, traveling up and over the berm that keeps out prying eyes. In the inner sanctum of the giant glass ring, he would stare at a pool of water that's been designed for being stared at. To go anywhere, he would apply his badge, which opens doors. Tired front-desk people, who drove in from Gilroy or Antioch (and are technically not quite tech workers or employees of the company at all), would say hello, maybe in recognition, maybe because that is their job. Work would then proceed.

At 9 p.m., he would return to the city on the private bus, disgorging into the Mission, having eaten three meals at work, gone to the gym at work, gotten a massage at work, spirited snacks out of work. To say that he lives in San Francisco is a partial truth. His mailing address bears a San Francisco zip code, but his Amazon packages get delivered to work, so they don't get stolen.

In some ways, San Francisco and Oakland, which increasingly play a similar role in the local economy as homes to tech workers, had improved for this tech worker. More and more restaurants catered directly and perfectly to him. Wherever he went, it was almost like déjà vu. He'd seen this place before, hadn't he? The aesthetic, its reclaimed wood, its swooping neon sign, its white tile, its heavy ceramics, its conception of a cocktail, its reliance on farro, its burrata oozing over this one farm's tomato.

The whole city ran on his money, readjusting itself to his needs like a flock of cybernetic starlings. What had been some inexplicable dollar store became a high-end Chinese place that would introduce him to baijiu, which would serve him well on his next recruiting trip to Shanghai. What had been a white-tablecloth establishment that made him feel a vague sense of adulting unease would be remodeled into a rough-hewn Italian place that serves Roman food, where a bone would emerge from the kitchen, and he'd eat the marrow ostentatiously, licking his fingers as the biggest red wine from Italy lay streaks on his glass.

Finding other people was easy enough. There were the work friends, and in the ring beyond them Hinge dates, and further afield, but as close to hand, Tinder people. Most services he might have sought out in the city—dentist, doctor, therapist—were available at work. Whatever wasn't could be accessed with a few taps of an app.

Somehow he always knew what his friends knew. They'd read the same things, sent the same tweets, bought the same sneakers, become interested in the same subcultures at precisely the same time. They could note the same trends in Uber drivers and home prices and venture capital firms.

It looked nothing like the conformist Levittown life of a young organization man in the 1950s, and yet . . . How could he go to any city in

America and have precisely the same coffee, leaning against the same wall, served by the same kind of tattooed person? There must be something other than this. The algorithms that fed him information must be subject to perturbation, to send him somewhere wrong. Isn't that part of what it is to live in a city, to have to negotiate people unlike oneself? Those people must be somewhere. How does one end up in the wrong part of town these days, anyway?

Only the homeless, sitting, sleeping, muttering, being, struck him as out of place. They were like a glitch, washed-up remnants of a city that's long gone. Bicycle tires, scrap metal, stray dogs, bags stuffed with plastic bottles. He tried not to look too closely, and then slipped into the bar, and the song was familiar. Friends crowded around someone's phone for one more Instagram story.

There's a certain kind of East Oakland street that's filled with tiny homes, built for the working class in the 1920s. House after house, stucco walls, tile roofs. Inside, two bedrooms, one bath, less than a thousand square feet. Kept up, the rounded windows and little details make these places feel cheery and bright. A fresh start out west for arriving workers! A starter home, back when that was a real idea in the Bay Area.

In early 2019, I found myself in front of just such a house on just such a block on the west side of 580 near Peralta Creek. Red roof tiles, a picture window edged in soft green, a Prius in the driveway.

This was where Brian McWilliams, once one of the most powerful labor leaders in America, had suggested we meet. He had been the president of the ILWU from 1994 to 2000. This was the very job held by Harry Bridges, the Australian leftist who led the union for decades.

This was a union that had been kicked out of the CIO in the midcentury for being too radical. While some locals discriminated, in San Francisco, the highest-ranking Black official, Bill Chester, recalled that the longshoremen of Local 10 helped open up other maritime unions, as well as acting as the political and sometimes physical muscle for the civil rights work of the mid-century. These days, the ILWU probably doesn't have a lot of real, actual Communists, but like an aging athlete, it never quite lost the build.

It was kind of the same with McWilliams himself, a bear of a man in a Hawaiian shirt. As he opened the door, he explained that this was his girlfriend's place. He'd gotten us sandwiches and beers, chips. I turned on the recorder and asked, "So, how'd you get into the union?"

"I was born in San Francisco," he began, which was not the full explanation, but was the enabling condition and perhaps the most important fact of his life. His father was an architect, and also an alcoholic. His family broke up and he ended up living in his father's office in the Tenderloin, a sleeping bag under a table for a bed. They had no refrigerator, so he kept his milk in the water tank of their toilet. Through a series of fortunate events, he ended up on the waterfront, making his way into the clerks, Local 34, who tracked the paperwork attached to the cargo.

He'd run with a whole bunch of San Francisco types in the late 1960s. They'd had some wild times! By contrast, the docks seemed tame. "I was bored to death on the waterfront," he told me. So McWilliams headed out to Japan. When he came back, he got involved in all kinds of local political activities. He created a business and folded it. But deny it as he might, the docks kept calling him back.

By the early 1980s, he was moving up the union local ranks, getting elected to different offices and joining the contract negotiating team that sat across the table from the Pacific Maritime Association, otherwise known as the employers. The PMA negotiators were often American shipowners, people who'd been in the maritime industry forever. "They were good people and, even the ones we didn't like, at least they were *our* assholes," McWilliams said. "Our counterparts at the PMA took their job seriously. They were professional people. We could talk, you know, and they were reasonable and we could work things out." They could make decisions about how the work would go and those agreements would stand without people across the world having a say.

But over the course of his career, things changed. As the ships got bigger and the cranes got bigger and the trade routes centralized, the people sitting across the table became an entirely new breed. "Now they're attorneys that are in for five or seven years or there's a turnover, there's a new corporate this or that," McWilliams said. "They're not shipowners. You're not sitting over there with one guy smoking a cigar and the other guy spitting tobacco juice in his cup. There are these guys in suits and ties, young people that are on their way somewhere. They're not going to be here in ten years. They're going to be somewhere else. They're not making decisions about the future. They're making decisions about their careers."

The ILWU managed through the brutal transition from break-bulk to containerized cargo in part because, as McWilliams put it, the employers "had an investment in us all being here for a long time." That might not be true anymore.

The whole value of the ILWU was the huge labor force that it could deploy to unload a coffee ship or get a vessel ready to sail for Hawaii, laden with all the goods the islands needed. Containerization changed the nature of the work. You didn't just need a strong back and a willingness to work. More and more, companies and workers prefer what are known as "steady men," guys who bypass the hiring hall. Worse, a lot of the work has simply been automated away with information technology or otherwise slipped beyond the jurisdiction of the ILWU. It's a familiar story for many kinds of working people, from manufacturing to newspapering to *Office Space*-style office jobs.

"So much of the traditional work that we've done has been replaced by electronic means, or it happens overseas or in all kinds of places, and we are being displaced," McWilliams lamented. "And instead of sitting across the table from people that would understand that commitment for all of us, we're sitting across from these attorneys that have not been there—or are not gonna be there—very long."

That hydra of young global lawyers is nearly as liquid as the capital that places them at those tables. It's not just the PMA, of course. Elites live in a global archipelago of cities that control where raw materials are extracted, how and where they are manufactured, and where they are shipped. An iPhone says it all: designed in Cupertino, manufactured in China from elements shipped in from around the world. That's a supply chain: a way of making products for a company, and also an ethics.

The simple answer for why the Bay Area got so expensive is that its companies used the Pacific Circuit to suck up money from all over the world and concentrate it here. And while the billionaires such as Marc Benioff or Tim Cook or Sergey Brin might *like* the Bay Area, that's a fundamentally different relationship from old wealthy shipowners who *needed* dockworkers to make their businesses run. In this rising tide of money, most people do not have a boat.

Given all that, ILWU leaders are not exactly apologetic that they have managed to create a high-wage union in a low-wage world. "We should be proud that we've survived and we're in the condition we're in with the kinds of benefits and salaries we have," McWilliams said. But he always wanted the ILWU to join the broader fight for everyday people, and not just here in the Bay Area.

In Seattle, during the anti-globalization protests at the World Trade Organization meeting in 1999, the ILWU shut down the port, and McWilliams gave a stirring speech. "When the ILWU boycotted cargo from El Salvador and apartheid South Africa, when we would not work scab grapes from the California valley or cross picket lines in support of the fired Liverpool dockers," McWilliams said, "these were concrete expressions of our understanding that the interests of working people transcend national and local boundaries, and that labor solidarity truly means that when necessary we will engage in concrete action."

McWilliams did not recite the basic litany of complaints about trade, nor extol the virtues of nationalism or wave the American flag in opposition to an interconnected world. "We demand fair trade—not free trade—not the policies of the WTO that are devastating workers everywhere and the planet that sustains us," he said. "And let us be clear. Let's not allow the free traders to paint us as isolationist anti-traders. We are for trade. Don't ever forget—it is the labor of working people that produces all the wealth."

And unlike many labor leaders, even today, McWilliams connected up the kind of justice that workers sought *with* the environmental concerns of many green activists in Seattle. The ILWU wanted "a world in which the interconnectedness of trade promotes peace and encourages healthy and environmentally sound and sustainable development," McWilliams said, "a world which promotes economic justice and social justice and environmental sanity."

The ILWU had organized not only for its own workers but for causes across the world. They'd done it so often and so effectively that the social scientists John Ahlquist and Margaret Levi wrote an entire book about the ILWU "to explain why some organizations expand their scope of action in ways that do not benefit their members directly." They called the book *In the Interest of Others*.

During the late '90s, culminating in that famous WTO protest, it seemed as if Huey Newton's ideas about revolutionary intercommunalism were coming to fruition. The internet had provided a space of solidarity for anti-globalization activists and they seized upon it. "Participants adopted a decentralized, network-based approach to political organizing and protest. Technological advances helped to open these new organizational possibilities," wrote the scholar Luis Fernandez. "Cell phones, e-mails, and listservs allowed for more grassroots, non-hierarchical modes of mobilization."

Subcomandante Marcos's Zapatista communiqués from Chiapas were read alongside Naomi Klein's *No Logo*. Mischievous urban culture jammers traded ideas with deep ecologists dedicated to protecting redwoods, farmers concerned about land reform swapped tactics with libertarian hackers.

This gave the movement a different flavor from the previous isms of the preceding decades. Leaders existed, but refused to be elevated into figureheads. For that brief moment, the technology provided a new, asymmetrical advantage to protesters and radicals. Organizing had always been so hard, and here was the perfect tool to help people *organize themselves*. "Thanks to the Net, mobilizations are able to unfold with sparse bureaucracy and minimal hierarchy," Klein wrote. "Forced consensus and labored manifestos are fading into the background, replaced instead by a culture of constant, loosely structured, and sometimes compulsive information swapping."

Perhaps the defining theoretical work of the time, Antonio Negri and Michael Hardt's *Empire* emanated straight from Italy's autonomist political scene, where Negri had been a leader. *Empire* fully formulated the economic and social changes that were taking hold as the Pacific Circuit grew.

It was the kind of theoretical book that crackled with the gnomic promise of a religious text. "In contrast to imperialism, Empire establishes no territorial center of power and does not rely on fixed boundaries or barriers," Hardt and Negri wrote. "It is a decentered and deterritorializing apparatus of rule that progressively incorporates the entire global realm within its open, expanding frontiers."

Here was a new vision in which Zapatistas on the internet and Malaysian Miss Free Trade Zone winners all made sense. "The spatial divisions of the three Worlds (First, Second, and Third) have been scrambled," the authors wrote, "so that we continually find the First World in the Third, the Third in the First, and the Second almost nowhere at all."

It transformed "Empire" from a single country's holdings to a U.S.-influenced network of states and other entities working for the same purpose: capitalist expansion. Negri and Hardt had something important to say about why cities felt so different. They were no longer centers of production but of *the control of production*. "As a mass demographic shift, then," they write, "the decline and evacuation of industrial cities has corresponded to the rise of global cities, or really cities of control."

The finger piers that surround San Francisco are now just office space for real estate investment trusts, AI companies, and yacht tour companies. What was the new Bay Area if not the most important center of the control of production?

Everybody goes to Target in the Bay Area, and so I found myself walking into the big-box store with a few sailors from the Philippines. I'd picked them up out on Middle Harbor Shoreline Road, right outside the terminal where their ship was anchored. They had a few hours, maybe half a day, and they wanted to be on land. Our meeting had been arranged by a local artist, Gabby Miller, who'd been on one of their previous voyages with them, making portraits of their families out of bunker fuel and videoing them playing basketball on one of the decks.

Her interest had been sparked by family history. Her mother is Vietnamese, her father American. They'd met in Vietnam during the war, and she'd never quite been able to reconcile how her family—her very life—was a product of violence, even if her father had been there as a humanitarian, not a marine.

Her research eventually led her, like me, to the Port of Oakland. During Occupy Oakland, she'd marched out there during a demonstration, and become obsessed. She found, of course, that 37 million tons of cargo passed through the Army Base and the container operations of Sea-Land on their way to Vietnam. The same war that birthed her family also birthed this new economic system.

Before the seafarers arrived, I'd been given a short list of items that one of the guys was looking for. Normally, journalistic detachment prevents you from helping anyone you're covering, but this felt different. They were risking their jobs to talk with me, as they had extremely strict orders never to talk with anyone from the media. Not only were they risking their current jobs but something like this could get them blackballed from future employment, too. Second, the list itself suggested a life, and a sweet one:

1. A pair of youth size 6 green Nikes (Kobes, to be precise).
2. Two singing Elsa dolls from the movie *Frozen*.
3. 1.7 ounces of Dolce & Gabbana cologne in the light blue variation.

The dolls and shoes were for little siblings. The D&G cologne, that was for the seafarer himself. I bought the stuff from Amazon, and when the appointed time came, I tossed the items in my trunk and headed for the port. I was a little early and the guys weren't off yet, so I drove down to the small ecumenical chapel that's located *right there* in the midst of the trucks and cranes: the International Maritime Center.

Inside, there were four clocks on the wall: Oakland, Manila, Hamburg, Mumbai. The IMC contained a little kitchen, some extremely international snack chips (shrimp-flavored corn chips, anyone?), a Ping-Pong table, and—of course—a tiny room for religious worship.

On the wall of that room, there was a poster of a multiracial ship crew, white man at the wheel, all watched over by a Jesus of infinite grace. Somehow, it suggests yacht rock, or at least the 1970s in a general sense. On the altar, the flowers are fake, but in the small book that seafarers use to actualize their prayers to God, there are real fears and hopes. Many of the messages ask God to protect their families, from whom they're separated. They thank Dios for their blessings, for making a hard job easier, for love, for their wives. Some messages are inscrutable: one asks for God's help with Ian, another thanks God for a wife and also asks for help with an annulment case. The names in the book suggest that people of many different cultures have passed through this door, stood before this altar, and asked their gods for help, before turning around and going to the big-box stores right on the border of West Oakland.

Because the truth is that the core activity of the International Maritime Center is probably not spiritual guidance but transport to the Emeryville shopping center. That's where Best Buy is. That's where Target is. And that's where the seafarers want to go.

Of course, they will take a ride, too, and I was offering. Three guys emerged from the terminal. The oldest I'll call O. He was quiet, white polo shirt tucked into hiked-up black pants, glasses. The youngest at twenty-eight, L, was fluent in English and, in most of his mannerisms and hoodie-and-Nikes style, indistinguishable from his American cohort. He worked in the kitchen. His boss was the third man I picked up, M, the ship's cook. He'd been working on ships for decades, and had been coming to Oakland since it was a dominant port on the West Coast, so he was dressed to blend in with the locals in an absolutely fresh Oakland Raiders vintage jacket. In all my photos from that night, a tiny ember catches the eye from the cigarette in his hand.

The guys had just run a standard shipping route. They loaded up in Yangshan, the deepwater port near Shanghai, headed across the Pacific to Southern California, where most import cargo goes, and now they were making a quick stop in Oakland to pick up food and wine and maybe some scrap paper. Once they left Oakland, they wouldn't see the west coast of North America for another six weeks. For them, it was like this for eight months at a go.

Despite our language failings, we were giddy as we walked through the doors of the Target. I had found seafarers willing to talk to me and they . . . were on land. In just a few hours, they'd be back on board and heading out past the Golden Gate.

M worked his way through Target to the electronics section. He was looking for a screen protector for his phone. I marveled at spending a few of your precious hours on land choosing or having to do such a mundane errand. But when you spend so much of your time out at sea, doing anything on land is pretty fun. "You've been there [on the ship] for how many months? You're just working twenty-four/seven," he told me. "You just meet your fellow crew every day, same place." He has four grown-up kids and he'd warned all of them off becoming seafarers. His three boys listened. His daughter did not. She was working passenger ships. One day, he hoped to open a restaurant back home, "an eatery or some kind of pizzeria."

Failing to find the correct screen protector in stock, M gave up and we went to grab some beer. I drove us up to Gabby's parents' house in the Berkeley hills. We hiked up to give the guys a view of the bay. Along the way, M spotted a bay laurel and snatched handfuls of leaves from its branches for his future adobos.

A bit breathless, we walked up a path and hit the lookout. Night had fallen. Lights twinkled in all directions, meridians from everywhere snaking toward the port and the Bay Bridge over to San Francisco. Floodlit cranes and ships stood against dark water. Across the Bay, a hint of Mount Tam's shouldered slopes. "Wow, amazing," was what M managed. What could be better than an unseasonably warm October night, just after sunset, looking down on this improbable, beautiful expanse? Were these hours seeing the world worth the months at sea seeing nothing but water?

The youngest of the three guys didn't seem to think so. He planned to get out. His father, too, had been a seafarer. Scholars and journalists sometimes call attention to the abuses that occur on these ships, and that have occurred on ships for centuries. It's real, especially in some countries and in some industries, such as commercial fishing. But, for most seafarers, the challenge is the banal brutality of the everyday work. "Once you're on a ship, you see how manual the labor is, how long and tedious and slow the movement of goods is across the Pacific Ocean," Charmaine Chua, a logistics scholar, told me.

L and I talked about his life. He'd grown up in a small town, and he'd signed up to work on a ship several years ago. His first gig was on a Korean car carrier out of Singapore. Walking up the gangway he told himself, "This is it. This is really it." But that very first day, when he got on board, the crew was gathered by the officers and told that there'd been a change of plans. Their new destination was Brazil, which would require forty-five days at sea.

"The first week, I am totally crying," L remembered. "I said, 'Oh Lord, please help me.' And I called my family and said, 'I want to go home.' They said, if you want to come home, it's up to you. And then I talked to the officers and they said, 'If you go home, you don't have money and you will [have to] pay for your ticket.'"

And he did need the money. He'd be able to save something like eight thousand dollars during his eight months away from his family and on the ship.

"I [was] just crying and crying," L remembered. The crew and the Filipino officer on board told him things would get better. "They told me they were like that before also." Maybe it didn't get better, but he got used to it. He tried to keep his mind there in the moment on the work at hand. And the job had perks: they'd see dolphins and whales out there in the expanse of the blue, though he never managed to get good video of them because from the great height of the cargo ship, even the biggest animals looked small.

I asked him about another perk I'd heard about: at night, in the middle of the Pacific, there's nothing around except dark water. The stars are as they would have been for our ancestors. All the hues and intensities of all those other suns just up there for your viewing, the Milky Way a river through the heavens.

At first, he nodded along to the idea that the stars were a good part of the job, then he stopped. "Actually, I don't like to do that, going out to look at the stars," he said. "Because [then] I remember my loved ones in the Philippines and then I'm getting sad. Yeah. You don't like to get sad because you miss your family and then you're going to cry again."

He didn't do the work for himself. As one of the older siblings of a big family, he was helping his siblings come up. His money was going to support his brother's schooling, and then when he finished, the money would go to help his sister graduate, too. Then he could think about himself, about getting married, about babies. "Kids are awesome," he said.

What kind of father would he be? Not that any of us can really answer that question, but he knew the kind he did *not* want to be. "I don't want to end up on the ship," he said. "I want to see my kids grow up."

I asked him about a guy like M, away from his family for so much of his life. "I experienced that," L said. "My father is also on the ship. We don't have a close relationship because he was always on the ship. Other children were close with their fathers. We are not close. Even now. Because when I'm on board, he is on vacation. When I am on vacation, he is on board. We don't see each other."

"So, when was the last time you saw your dad?" I asked.

He thought. "Five . . . years ago?"

We wandered back down the hill. I handed over the Nikes, the singing Elsa dolls, and the D&G cologne. He handed me back the precise number of dollars they had cost. I thought about the millions of dollars' of fuel his ship would burn, the hundreds of millions of dollars' of goods onboard, the debt and derivatives sitting on top of all these transactions, these goods, the real estate all around us. L's story was one of hundreds of thousands.

I drove the guys back down to the port, which was still humming. Everything felt high-contrast—floodlight and pitch-black. M walked through the parking lot to their ship, visible until the very last by the silver of the RAIDERS on the velvet black of his jacket, bay leaves tucked in one pocket, a dream about opening a pizzeria in the other.

The system, working as intended, simply asked people to sacrifice so much. You gave up everything most of the time to have a little something sometimes.

24

Transpandemic

A transapocalypse is a spectrum. Some parts of the world will experience death and suffering and tragic upheavals as horrible as any humanity has ever seen, even while others experience unprecedented prosperity.

—ALEX STEFFEN

In retrospect, 2019 was the year when the arc of Oakland's economic and cultural ascent began to wobble. Perhaps it was purely symbolic, but the Golden State Warriors completed their move to San Francisco, losing the NBA Finals that year in grotesque, haunted fashion. Perhaps it was possible that a town like Oakland didn't *need* the Warriors anymore. Oakland was a new place.

But the Warriors were one of the few things that knit old and new Oakland together. You could be standing anywhere, doing anything, in 2015, 2016, and someone would strike up a conversation about Steph and Draymond. There was conflict, but on that one thing, we could agree. I watched many playoff games at a little spot called Oakland Hot Plate. I'd gone to the championship parade in 2017, when a million people crammed onto the streets of downtown, blue and yellow everywhere. A young man kept shouting next to me, "Look at a million people in my town right now! This is my town!" to no one in particular. "A million people! And we all love each other. It's all love. All love!" Others just shouted out their neighborhood: "West Oakland!" As Steph's float went by, one old cholo hollered, "He's a baaaaaad man."

It was so beautiful, feeling like everyone was in it together, if only for this one stupid sport and team that we loved. That evening, I went to an event with Margaret celebrating the Urban Strategies Council's thirtieth anniversary.

With the commanding and elegant Angela Glover Blackwell—Fred Blackwell's mother and founder of the council—onstage, Margaret rolled around in her Warriors jersey, talking about a new bill that had passed the California legislature that might allow West Oakland to develop its own clean air action plan.

At that Urban Strategies Council event, someone asked, How are we gonna keep Oakland Oakland, basically, with all this new money coming in? "If we don't figure out a way to preserve affordability, this community is going to lose it all. And this is a great community because of its diversity," Glover Blackwell said. "My daughter went to the parade and she was showing me photos. I looked at those one point five million people. It was wonderful. And she said, you know what's so great about Oakland's diversity, it's that everyone interacts with each other. Oakland is such a special place. And the advocacy is part of what makes it special. It's no surprise that Oakland is the birthplace of the Black Panther Party."

For a time, I wanted to believe it was inevitable that Oakland would put it together. That we'd show the way to create a diverse, mixed-income city. If any city in America had its politics aligned right to try to accommodate an influx of Pacific Circuit money, it had to be Oakland.

But that's not what happened.

Before the pandemic shutdown, the last reporting I did was on a meeting of the San Francisco Small Business Commission on the topic of "ghost kitchens," a newish concept popularized by the deposed Uber CEO Travis Kalanick.

The logic went like this: if the problem for restaurant businesses was maintaining a brick-and-mortar location with front-of-house staff, then why not just cut out those costs? You could *look like a restaurant* on a delivery app, without maintaining a building where someone could grab a drink while she waited for a friend, or take up a crappy table while he read a book. What ghost kitchens *did* do was compete with the businesses who had that whole other cost structure, which just so happened to provide a bunch of positive externalities that were not captured by the restaurant itself. The whole idea was a fine startup, I guess, but it was *ruthless*, and San Francisco restaurateurs were up in arms. They hated the idea. And Kalanick, too.

That was February 20, 2020. The "novel coronavirus" was a Chinese phenomenon. And during that period, many people, myself included, had the same wrong thought about where COVID-19's impact would be greatest: supply chains. "Suddenly, all supply chains seem vulnerable because so many Chinese supply chains within supply chains within supply chains rely on each other for parts and raw materials," Rosemary Coates, a supply-chain consultant, wrote in the trade journal *Supply Chain Management Review*. "That tiny valve that is inside a motor that you are sourcing for your U.S.-made product is made in China. So are the rare earth elements you require to manufacture magnets and electronics."

In those early days of February and March, as corporate and consumer confidence plummeted and the Great Stay at Home began, supply chains were a mess. Seventy-five percent of businesses reported problems. Prologis recorded the sharpest decline in its customer sentiment survey ever. The people who imported and exported goods were as worried as they'd been in the era of globalization. It was the most serious ever threat to the Pacific Circuit.

During the panic weeks before the long doldrums of the pandemic, toilet paper seemed to be the item in short supply. But toilet paper is actually manufactured here in the USA. Warehouses and trucks were not empty, but people's demand for toilet paper skyrocketed and cleared out the distributors. Once panic buying slowed, the shelves were quickly refilled.

A better example of the deeper supply chain problems came in semiconductors. American firms maintain more than 60 percent market share in the massive overall business by controlling production, but more than 80 percent of manufacturing is now done in Asia. A single company—Taiwan Semiconductor Manufacturing Co. (TSMC)—makes more than half of all the silicon wafers. After that, it's Korea's Samsung, then Taiwan's United Microelectronics Corp.

Manufacturing in Asia grew up *with* Silicon Valley. And the Pacific Circuit has some strange effects that are not quite intuitive. While global supply chains produce an astonishing array of products, the tendency within different industries—absent strong state or international law—is toward massive concentration in a given function of the system. The benefits of scale and specialization are hard to counter.

Semiconductor production sits at the bottom of the entire world of products because transistors are the basic component of all kinds of computer chips. Semiconductor fabrication requires huge factories called "fabs" that refine and produce huge cylinders of silicon ingot, then slice them into thin wafers, then mask and etch them with complex circuit designs.

Every step—and countless ones that I've glossed over—has been relentlessly improved by the whole industry for fifty-plus years now. As you might imagine, manufacturing the most pure chemicals ever known on earth in an environment that's much, much, much cleaner than the cleanest hospital operating room so that you can create tiny, complicated devices requires a *lot* of money. Traditionally, a new process size—think of it as a generation—gets commercialized every two or three years (for example, 350 nm in 1993 to 250 nm in 1996 to 180 nm in 1999). To make the new semis, you have to build an entire new factory, get it up to speed, and soak up as much profit as possible before the cycle starts over. This is how computers have gotten faster and cheaper.

There is a much lesser known corollary of Moore's Law about the improvement of computing called Rock's Law, which states that the cost of a new semiconductor foundry doubles every four years. Rock proposed that just a few companies might be able to maintain vertical integration—that is, both designing and manufacturing chips. The rest of the companies would end up specializing. And that's what happened.

As the 1990s turned to the 2000s, Taiwan Semiconductor Manufacturing Co. began to dominate the business. As American companies prospered by designing and branding chips, TSMC steadily grew its market share, so that by the 2020s, much of the world's semiconductor production took place near Hsinchu in Taiwan, a seismically active region in perhaps the most geopolitically disputed territory on earth. Oops!

Concentration creates risk, and the flows of the Pacific Circuit create concentration. And it's not just chips, either, but all kinds of stuff. "Today we see such concentration in the production of many if not most of the components that go into computers and other electronics," the anti-monopoly researcher Barry Lynn said in testimony to a Senate subcommittee, "but also in products ranging from pharmaceutical ingredients to vitamin C to piston rings to pesticides to silicon ingots."

Competition and concentration have created incredibly low prices. There is breathtaking innovation and profit at the leading edge. These businesses are lean and mean. They redistribute a lot of profits to investors rather than reinvest in long-term capacity. Other countries have provided massive support to their domestic semiconductor industries, whereas the U.S. government, in thrall to the idea that the market would solve all problems, did not. But as long as the system remained in balance, cheap chips flowed out of Asia and into the world's products, from microwaves to cars to little beeping toys for children.

When COVID hit, suppliers *did not* shut down production. Stricter COVID protocols combined with special carve-outs for the industry in Asia largely kept the fabs running. Rather, as the U.S. economy nosedived, automakers pulled back their orders for the old, cheap chips they needed to work power windows or control the screens in their cars. Naturally, the few companies capable of making the older chips responded by selling or promising to sell those goods to other companies. When both electronics and car demand came roaring back, as they did all over the world, the automakers were out of luck. The chips had already been sold.

The problem is not easy to solve. An increase in demand, if there's no capacity, simply means fewer people get the chips they need. What would it take to build capacity? Well, as described earlier, you'd need to build a *node* that could produce the older chips. The Commerce Department commissioned a report on where the problems for American manufacturers lay and it was at the 40, 90, 150, 180, and 250 nm sizes. The 40 nm process was commercialized during the Bush administration. The 250 nm process in 1996! No one wants to build massively expensive facilities for an ancient technology. And even if they could, where would they get the know-how and the equipment?

There just was no slack in this highly concentrated, highly efficient system. And so: there was a shortage that took years to work out. Carmakers reported they forewent sales of more than $200 billion (!) because of the lack of chips.

With booming demand and a constrained supply of cars, the market *did* actually do what would be expected, and prices for cars rose. White House economists blamed cars alone for more than a third of the inflation that began to dog the U.S. economy in 2021. Writing that summer, they noted that cars are "the industry of industries" and that the prices of thirty thousand parts affected prices of cars. But it was the semiconductors, the parts of a car most closely connected to the transpacific tech circuit, that did the most damage.

As the supply chain specialist Hassan Khan put it, "The weakest link in the supply chain is often an inexpensive part." That's because inexpensive for a buyer often means low margin for a seller. The little things are taken for granted in normal times, but also subject to brutal competition with razor-thin margins on the outskirts of the mainline big business.

This was true even for the early U.S. response to the pandemic. Testing was hobbled by the lack of pipettes (and chemicals and vials). Our mitigation measures were hobbled by the lack of masks. States battled one another for PPE for their hospital workers as national coordination completely broke down, and the standard supply chains for N95 masks gave way to late-night airport trips and shady deals.

A major part of the early miserable U.S. government response to COVID was the lack of domestic manufacturing of important products, the resulting dependence on global supply chains, and the arrogance of imagining that the United States would somehow come first in a marketized world we'd asked for and built.

Taiwanese semi firms, Malaysian assemblers, and Chinese mask manufacturers had no ethical responsibility to change how the Pacific Circuit functioned. By the very logic of the system, in fact, they should have extracted the highest possible prices without regard for the health and safety of the American people or economy, and they did.

At home, San Francisco's downtown emptied. Oakland, too. San Jose, the same. On the street level, trash scudded in the wind like tumbleweeds. Tents dotted sidewalks that were once bustling with, like, Goldman Sachs executives. These were not ghost towns, these were ghost cities. Everyone alive knows what a strange time it was, the new weird rituals, the way it felt safer to stay in your house, or if not there, then your block, maybe a loop around the neighborhood.

Societal forces that had felt like *gravity* were revealed to be . . . nothing. Your kids didn't have to go to school. You didn't have to go to the office. No one had to go to restaurants. The government could simply send everyone money. They literally could just do that.

In the early months, I was deeply enmeshed in efforts to get good data around the pandemic. I saw hundreds of people sweep into the pop-up organization I co-founded at *The Atlantic*, the COVID Tracking Project, which supplied data to hundreds of media outlets and many millions of people. My neighbors pulled together into a pod before those even had a name. I thought of those women on the Richmond home front, going to work in the shipyards and then taking care of one another's kids and families. Was this disease going to be the thing that brought Americans together, that saw us extending our circles of care, that showed us a new way out of whatever version of capitalism we'd found ourselves toiling within? Perhaps American society's entire value structure would shift, placing a premium on essential care work and community building. As bad as those early months were, as many people as were dying, as listless as the days felt, hope fluttered in my heart as I watched the people around me just *trying so goddamn hard* to make this new weird life work.

A phrase from the historian Richard White kept running through my head. In his work on the late nineteenth century, when industrialism was at its most vicious and wealth inequality was spinning out of control, new cities like Chicago were ravaged by disasters constantly. Fires swept from rich to poor and poor to rich. Waterborne illness circulated throughout the city, with no regard for class distinction. The labor problem seemed intractable; housing killed people. What emerged from that hellscape was what White called the "democracy of defecation"—a rugged realization that a city needed a basic level of infrastructure that benefited all. Water, fire, sanitation, transportation: these systems needed to cover everyone to truly protect anyone. "Cities were like ships," White wrote. "They sailed, and sank, as a whole."

That had to happen, I thought. We had to extend the circle of care beyond ourselves—far beyond—or we were all gonna sink together.

I probably don't have to tell you that cities, especially those in the Bay Area, did not sail. They sank.

After the early fears that the Bay Area's real estate market would implode, prices soared. Tech workers did just fine raking in money from newly kitted-out home offices. The residential real estate spiraled upward, and though it eventually moderated, it was a sign that the city was fracturing.

People whose jobs were connected with the Pacific Circuit turned out to do extremely well. Billionaires increased their wealth more in those early years than was possible to comprehend. The rich got so much richer, and many, many, many people working at the local level have never recovered.

In my talks with economists, it became clear that there is a massive shift going on in the basics of how our urban economic engines are running, on par with the shift brought on by containerization. But no one is quite sure what it is or what to do about it. The pieces of the city don't fit together like they used to, yet we have urban plans and ossified governance structures that make change slow, slow, slow. We already jerry-rigged a new way of living atop the old industrial city skeleton, but now the city that we're seeing take shape has even less concordance with the nineteenth-century political units that we're using to govern these places.

The old industrial cities did the work of city life. They made products that were used right there. People lived in them and worked in them and consumed in them. The rich and the poor were right there all together, riding the cable car. To improve the city, one had to improve the underlying infrastructure. Put in a new transportation system, build a new factory, add port facilities, increase the capacity of the government, construct ten thousand small houses for new residents.

There was an incredible integration to these cities. The old industrial city, a place like Oakland, really did need worker housing and public transport.

By contrast, the subsidized construction of the suburbs weakened the cohesion of the American city, as did the changing economic geographies brought on by containerization. Manufacturing facilities could be built anywhere, from exurb to Asia. Then came the big-box stores like Target and Walmart that robbed the downtowns of their retail power.

Cities survived all that, though. Through the early twenty-first century, millennial demand began to return to cities, delivering life to inner cities that had been struggling for decades.

For a time, it kind of worked, in a dispiriting way. Cities became a place where people who stared at screens paid everyone else to service and entertain them. Uber and DoorDash and all the rest of the personal logistics companies grew into that space. People who tapped the Pacific Circuit had no time, but money to burn. Everyone else was desperately hanging on and willing to do whatever those rich people needed. I ended up calling it the "servant economy."

People's consumption used to leak out into the local areas where they lived and worked. That's the whole idea when people say one tech job creates X other jobs, which is a common reason that cities around the country try to cultivate local high-technology industries. It's also an idea often spoken aloud that Black people should buy Black and "circulate our dollars within the community."

The ghost city that's rounding into shape after the pandemic militates against the spreading of the wealth. Rich people's consumption can be perfectly precise. They can cut out almost all the locals hoping to grab some dollars from the global fountain. Anybody with a screen can just buy the things they're targeted with on Instagram, which get shipped directly from China to their door. With commercial rents going up and housing being so expensive for their workers, many local spots have simply been wiped out.

The societal fabric, a thing I was never quite sure existed, tore. You saw it in the smaller things—like people driving totally crazily or a step-change increase in chronic school absenteeism. You saw it in bigger things—like Oakland's murder rate soaring back to near crack-epidemic levels or homeless encampments growing into huge informal settlements like the one at Wood Street, at the edge of West Oakland. You saw it in new things—like fentanyl killing hundreds and hundreds of Bay Area residents per year, as they got high on deadly mixes of the opiate and meth. People are taking so much fentanyl that a paramedic told me they used to give people ODing on heroin less than one dose of Narcan, a drug that reverses an overdose. Now they might have to give five doses to revive someone. That's how much pain people are trying to kill.

Oakland's wealthy residents now draw on a labor shed of workers from all the way in Sacramento, often Black folks who used to live here before they were displaced. They return now, as delivery workers or Uber drivers, to neighborhoods they grew up in, shaking their heads at the changes that have occurred. Then they disappear back to new cities and towns, knowing they'll almost certainly never be able to afford to come back as residents to the places where their ancestors lived. These past residents exist as traces on a Ring doorbell, a sign for a restaurant that closed, a small office filled with empty desks, a receptionist's old calendar still turned to March 2020, the moment the old city officially died.

During one of those long pandemic afternoons, I took my kids down to a beach in Alameda. The sun was shining, the water blue. I'd been there dozens of times in the twenty-teens looking at shorebirds, playing along the edge of the water, digging around. Under normal circumstances, out in the bay, there might be a sailboat or two, some paddleboarders. But on that day in February 2021, there was a row of container ships, the huge letters of their corporate parents visible across the water—COSCO SHIPPING, CMA CGM, HAMBURG SUD. Having spent a lot of time on the bay's shoreline, I was shocked. What were these ships doing there? These were the secrets of the port, surfaced there in plain sight.

What I saw was just an echo of the real problem emanating from Southern California. There was an epic, globe-altering traffic jam at the Los Angeles/Long Beach port complex. And, like many traffic jams, this one had multiple causes, some chronic and others quite acute.

The pandemic sequestered people in their homes. Even if they went out, many of the services they would have spent on—restaurants, travel, concerts, sporting events—were not available or desirable. And at first, Americans were not experiencing unusual economic distress. In fact, quite the opposite. Thanks to solid government intervention and the lack of places to spend money, Americans began stockpiling cash. For years, people were saving just a bit of their disposable income—6, 7, 8 percent. The measure had gone over 15 percent only once since 1960. In April 2020, the savings rate hit *33.8 percent*, almost double the highest monthly number ever recorded. That's how unprecedented the beginning of COVID times was.

As the early restrictions eased, Americans began to spend down some of those accumulated dollars, but there was a *lot* of excess money sloshing around in bank accounts. With money to burn and some services off the table, Americans started buying more goods. Those goods, by and large, come via the Pacific Circuit, which meant that they had to be shipped to the United States. Starting in August 2020, the backup began to build. The average shipping time from China to the U.S. climbed from about forty-five days to sixty by February 2021, when I saw the ships anchored in the bay, and eighty by the end of the year. The delays were largely at the American end, where ships were spending as much time anchored off the California coast as they did transiting the Pacific. By early 2022, ships were waiting almost eighteen days to unload.

With so much demand, the cost ballooned, too. Containerization had driven transpacific shipping costs down *so much* that they were barely a factor in the economics of a lot of products. Historically, to ship a forty-foot container from China to the United States cost $1,800–$2,000, but the price had been heading down and varied depending on how much volume an importer could deliver. For a huge importer like Walmart or Target, the price could be as low as $800 per container. By February 2021, the price had crossed $4,000. And then, as summer arrived, it exploded, more than doubling from that point as retailers became willing to pay whatever it took to get their goods. In the fall holiday shipping season of 2021, it cost more than $11,000 to ship a container, an 8,500 percent increase from pre-holiday norms, according to one shipping data firm.

On the one hand, this has been, by far, the biggest trouble in the Pacific Circuit since the creation of the system itself in the 1960s. On the other hand, the shipping companies have *made a killing*, pulling in something like $150 billion in profits in 2021, enough to erase the huge losses of previous years. (In 2016, for example, the shipping industry was losing something like $10 billion a year.)

As the system's flow stopped up, key elements also got out of balance on both sides of the Pacific. John McLaurin, president of the Pacific Merchants Shipping Association (PMSA), told me the retailers he represented were struggling with every step of the journey. "When they've got product in China, they can't get a box. When they get a box, they can't get a ship. When they get a ship, they can't get a berth," McLaurin said. "When they get a berth, they can't get a truck. When they get a truck, they can't get a chassis. And when they get a chassis, they can't find a warehouse to take it to."

With so many boxes coming in on the import side, there were not enough empty containers in China, despite the many empties that are shipped back across the ocean. The import-export imbalance has been a structural feature of the Pacific Circuit since the very first containerization systems were imagined, but it has usually been manageable. Not so during the pandemic. So, no boxes.

Shipping, up until the pandemic, had seen incredible consolidation. The large shipping lines—especially the biggest, Maersk—have led the industry to be remarkably concentrated in just a few firms. Before the pandemic, the shipping industry had not seemed poised for remarkable growth. Its planning stretches years into the future. There was no way to simply throw more ships onto the water to soak up the increased volume. So, no ships.

Centralization rules the port business, as it does shipping and semiconductors. Since the 1980s, the complex at Los Angeles/Long Beach has eaten more and more of the container market. It now moves 20 million twenty-foot equivalents (TEU), the measure of container traffic. Why couldn't importers simply send more boxes through Oakland or Seattle/Tacoma or San Diego? By the PMSA's reckoning, just the increase in imports from 2019 to 2021 in Southern California—1.6 million—is a substantial percentage of the total imports that the rest of the West Coast managed, 2.7 million. Oakland's total import volume was a record, and just barely crossed 1 million.

Ports, both Asia's export giants and America's import giants, cannot add capacity easily. A modern container berth requires huge amounts of shoreline space, massive specialized equipment manufactured by Asian companies such as ZPMC, and a whole ecosystem of niche maritime companies. In each major port, there's an institution, usually called the Marine Exchange, that provides information to the myriad companies that work the waters of each place. The marine exchanges publish these big, fat books that people can use to find a barge or a tugboat or drayage truckers or (really) a priest for the crew.

And the problem with expanding port capacity is that you've got to expand all these services, too. The whole ecosystem—almost like a semiconductor node—has to scale up. But no one wants to own the unprofitable, difficult parts of the system. No one wants their core competency to be a complex yet low-margin business. Thus as in the older semiconductor business, so it is with the ports, and the symptom of that problem is the humble chassis.

In order to get hooked up to a truck, a container must be loaded atop a chassis. Managing these chassis is done in different ways in different countries and ports, but in the United States it had long been the shipping companies' responsibility. This was simply a historical holdover from the way U.S. container shipping developed, within vertically integrated companies such as Sea-Land and Matson. The rest of the world might place the responsibility on truckers, or freight forwarders, or the people importing goods.

In 2009, Maersk simply said it would no longer be vertically integrated, and spun out a chassis business. The whole industry followed suit, and the result has been "a convoluted patchwork of arrangements that vary from region to region and has often struggled to meet the goal of providing a ready, easily accessible supply of chassis," as the shipping industry bible, *The Journal of Commerce*, put it.

In 2016, back when the shipping industry was deep in the red, the Korean line Hanjin went out of business. It was a disaster. Thousands of empty Hanjin containers sat atop chassis all along the West Coast. They were stranded, more or less, and that kinked the whole fragile flow of chassis through the system. To solve the problem, the Port of Oakland itself stepped in and provided storage for empty Hanjin containers. Vast stacks of blue Hanjin boxes piled up in a storage yard. Eventually, things evened out, but it was a foreshadowing of what would happen in 2021. Amid the surge in imports, the chassis got imbalanced down in Los Angeles/Long Beach. All the links that had made logistics smooth were gumming up, and gumming up one gummed up the others. So, no chassis.

If chassis are scarce, truckers have to work harder to find one. Waiting in two lines of trucks, not just one. If a warehouse is short on workers or a load isn't ready, they wait again. It's a pure loss. All that added to the long-term problems of the truck-driving business. Low margin, physically brutal, unpredictable, precarious: drayage trucking is the kind of working-class job you find there at the end of the race to the bottom.

Despite the fact that there are many more Californians who *could* drive a drayage truck and go pick up boxes in Long Beach and Los Angeles, not enough truck drivers could be found. This is no surprise. As we saw across the American economy in 2021, there were many job openings—many millions—that simply could not be filled because no one wanted that gig. And actually, because the truckers are independent operators, there's not even a job to fill, just an unprofitable, small-time niche in the vast web of global trade. So, no trucks.

Add it all up and the smooth flows of lean manufacturing and just-in-time deliveries shuddered along, taking products off shelves, frustrating retailers, and enriching a few companies that were well positioned to take advantage of the new conditions.

The problems got so bad that the Biden administration stepped in to try to fix things at the LA/Long Beach port complex with an executive order in October 2021. The main thrust of the change was to move the port complex to 24/7 operations, but that would require all the players in the supply chain to make that move. ILWU workers, with high salaries and long careers, might be willing to take those nighttime shifts. But what incentive is there for a drayage trucker, who might not even be driving a big rig in a year, to move to that schedule, making an already terrible job worse?

Watching the American state struggle to contain COVID, and then flounder trying to smooth out the kinks in the supply chains of the West Coast, proved how much our government has been hollowed out. The actual capacity of the American state to do *anything* but pump dollars through the financial system is a big question. Even when the ambition might be there—as in the semiconductor industry—the Biden administration could offer only longer-term solutions and giveaways to American companies. On the port side, the Biden administration could offer "leadership" in meetings between a dizzying array of companies and governmental entities. What it could not do was actually fix the problems with government resources and staff, nor force private companies to change their existing fragile long-term practices.

It is difficult to peg precisely how these supply chain problems were stoking the rise in prices that we call inflation. There were so many factors stoking the U.S. economy in 2021 that it was hard to know where to pin the blame. But most analysts agree that it was a good chunk, probably enough to have severely and perhaps permanently weakened the Biden presidency.

25

The Grand Bargain

I parked near DeFremery Park in West Oakland on a wet, cold December night. It was that sloped-shoulder time of the year, when the sun seems to go down shortly after it rises. A couple guys were sitting in a lived-in car along the road, smoking, talking, staying out of the rain. The streets were puddled.

Ms. Margaret and Brian had a big event: the five-year celebration of their work on a controversial environmental justice bill called AB 617. The 2017 bill funneled some money from California's greenhouse gas cap-and-trade system into helping local neighborhoods develop community emission-reduction plans. Basically, money that is generated by polluters gets plunked into a special state account called the Greenhouse Gas Reduction Fund. And some of it is directed to a dozen "environmental justice communities" around the state, places like West Oakland or the neighborhoods around the LA/Long Beach logistics cluster.

While national organizations such as the Brookings Institution saw a national model in the bill, some environmental justice groups *hated* it. The California Environmental Justice Alliance wrote a scathing report about its failings. That coalition of EJ groups says the bill wasn't a good "compromise," as its backers contended, but merely a "political ploy to advance the cap-and-trade agenda." Some local orgs straight-up refused to participate in the program.

Because they'd been collecting data for years, both through official channels and in other efforts like the Google Street View cars, WOEIP could skip some of the early stages and get right to working on a plan in West Oakland. That is, if they wanted to do it. "In the end, our organization has always said, 'What's the opportunity?' We do not like the roots of the opportunity, but there's money on the table and we would be stupid to walk away from this. And that was ultimately the decision we needed to make," Brian said. "Even though we had allies who were going, 'That's tainted money. You got to reject that.' We couldn't see how our community would get better if we walked away from the financial opportunity that was created by the legislature under cap-and-trade."

By the end of 2023, tens of millions of dollars had gone into West Oakland to develop and implement the plan, a small sliver of which has gone to WOEIP. Along with settlement monies received from a lawsuit against Schnitzer Steel, the organization now has twenty times more dollars flowing through it than when I first met Margaret. It may only be a couple million dollars, and Margaret may still make $42,000 a year, but it's a massive change in the capacity of the organization.

In exchange, they had to build the whole community alliance out and then get it to actually function to create an emissions reduction plan. They began work in 2018, and on that cold December night, they were celebrating half a decade of coming together through the pandemic era's wild dislocations, through the spiking violence in Oakland, through the side-eye from some other EJ groups, through the whole Tagami saga.

Those five years were an eternity in the micropolitics of West Oakland.

When they began, Libby Schaaf was mayor and Oakland was booming. By that December event, a new young mayor, Sheng Thao, had been elected. Supporters of the export terminal, including a private equity firm with a stake in it, funneled hundreds of thousands of dollars to Ignacio de la Fuente, a former head of the city council, Tagami ally, and Margaret opponent. The funds represented the bulk of de la Fuente's campaign financing. It was a seemingly brazen attempt to buy the Oakland mayor's race on behalf of coal interests. It failed miserably. De la Fuente's campaign never took off and he finished a distant third.

Thao's parents had been Hmong refugees fleeing the wars of Southeast Asia. She'd grown up in Stockton's very tough public housing projects and left the traumas of home to eke out a solo teenage existence in Richmond. She found herself the victim of domestic violence. For a time, she lived in her car with her infant son. But she got her life on track, went to Merritt College (same as Margaret), and worked her way up through City Hall.

Their councilwoman in West Oakland changed, too. Lynette McElhaney had been challenged in 2016 by Noni Session, the same Noni who had grown up in the shadow of the Cypress Structure. After the earthquake, she'd gone on to a BA from San Francisco State, then doctoral work at Cornell in cultural anthropology, mentored by the radical anthropologist Terry Turner. When she came back to West Oakland, she was fired up by the changes she saw in the neighborhood. McElhaney, she said, was "throwing the doors, or keeping the doors open, for unchecked speculative development in West Oakland," which served as a nexus for San Francisco, Silicon Valley, and the East Bay.

As her run got off the ground, her campaign manager, the organizer Carroll Fife, began introducing Noni to her contacts. "And Margaret was the first one," Noni remembered. "She received me at her house. For three solid hours, we drank red Kool-Aid and vodka, and she gave me a detailed rundown of the arc of the port, the people involved in its consistent obfuscation of its responsibility. Then they put me in the room with Brian Beveridge, where he went through all the energy policy with me." After those meetings and many like them, she was able to run an impressive grassroots campaign. She raised only $10,000 dollars, but got 43 percent of the vote. She still lost, but it was an impressive loss. There were new political alliances growing in West Oakland, powered by the OGs' knowledge and the energy and sophistication of the new generation of leaders.

Most electrifying, on November 18, 2019, a group of Black homeless mothers moved into a home owned by a subsidiary of the real estate investment firm Wedgewood. They were, they knew, trespassing on the property, but the occupation had an immediate moral drama that the media could not resist. After all, the property was vacant. And here were mothers of small children who were living on the street. Wasn't sheltering them the highest and best use of the property?

The property itself, as the journalist Katie Ferrari excavated, bore all the marks of racial capitalism. A Japanese family, the Nakamuras, lived there. Then they were incarcerated during World War II, and the house was sold to a Black couple, Carter and Willia Mae Roberts. They held on in the house until 1987, when Willia Mae died. An investment company bought the house and resold it. One owner, Betty Mack, bought the property with a subprime loan, and ended up among the foreclosed in the late '90s. Eventually, it found its way to the Wedgewood subsidiary in 2019, which specialized in buying "distressed" properties and flipping them. And it would have done just that had Moms 4 Housing not come on the scene.

Attention flooded to the mothers, and one person emerged as their most influential spokesperson: Carroll Fife. The Moms 4 Housing occupation lasted an action-packed two months until sheriffs evicted the mothers, but in the meantime, they were able to get Wedgewood to sell the house to the Oakland Community Land Trust, taking it out of the market permanently.

Moms 4 Housing inspired copycat groups across the country and garnered international coverage. Throughout the occupation, Fife had drawn attention to the dual nature of housing as both a human need— or in her frame, a human right—and a financial asset floating in an ever more global and financialized market. Fife rode the organizing wave to a big win in the 2020 city council election, trouncing McElhaney.

The big WOEIP event was held inside the community center at DeFremery. Outside, they'd parked a huge demonstration semitruck that used hydrogen instead of diesel. All the organizations involved—nonprofits, the port, planners, health advocates—had arranged themselves behind folding tables along the walls. Chairs filled the center of the echoey room, facing a stage two steps up from the floor.

The crowd was raucous and talkative. These people knew one another! All kinds of movers and shakers filtered into the room. For several minutes, Brian and other staffers tried to get people to quiet down and take their seats. They were unsuccessful. Then Margaret let out one great yell, and that did it.

Brian spoke, calling their work on the air quality improvement plan "an amazing experiment in community power." He told the story of how the work had begun in confusion. Whatever the legislation said, they had to figure out what a community air plan was and how it should work. "What's the first chapter?" he said. "What comes after that?" The first chapter, he noted, came out of the stories community members told in their meeting. It was then drafted by Bay Area Air Quality Management District staff. "And it pulls no punches about racist land practices, redlining, destructive historical decision making by government agencies," he said. "Those don't amaze me. But it amazes me that a government agency could recognize them." Of course, it helps that BAAQMD had a heavy-hitting EJ person in place already, the lawyer Veronica Eady.

But they needed the rest of the plan, too. The California Air Resources Board gave them one year. The community WOEIP pulled together met once a month, and never had fewer than seventy people in the room. As the official partners on the plan with the air quality agency, Margaret and Brian met every goddamn week with the BAAQMD.

Margaret saw all their work as being about relationships. "If you don't do shit together," she told me, "how you gonna know about the real shit?" They went way back with BAAQMD. In 2009, they did a truck survey. "The engineering and science and technical staff walked with us in the neighborhood. Don't that make a difference?" she asked. "And as they did calculations on what we found, they taught us what that looked like."

Ten years later, they were working with BAAQMD on the AB 617 plan, which moved to the implementation phase in 2020. EJ groups had specifically targeted implementation as the weak spot in the legislation. Part of the compromise of the legislation, as it often is, was that there weren't strong enforcement provisions to cut emissions from tailpipes or smokestacks.

The West Oakland working group found that the legislation created some space for the plan to "reduce exposure," not just emissions. And when they looked around, they saw a whole bunch of government agencies that could reduce exposure. No one said that a community couldn't simply talk government agencies into reducing their emissions. "City governments can reduce exposure because they control land and traffic and zoning," Brian told the group. "[Oakland] now has twenty strategies in the plan." The same went for businesses and multiple other agencies. In the end, they put together ninety strategies with different partners. Beaming with pride, Brian told the group, "We're in year four and about ninety percent of those strategies are in implementation in one way or another."

In effect, Brian and Margaret had taken legislation that just about everyone agrees is broken and hard to use to help a community, and created a radical platform to improve air quality in West Oakland, funding local people all along the way.

Margaret introduced Cristina Garcia, the author of AB 617, to the crowd. Garcia's time in the legislature had been rocked by a sexual misconduct scandal, shortly after AB 617 passed, but she was also from Bell Gardens, a corner of Southern California that happens to be just up the 710 freeway from the LA/Long Beach port complex.

"I've been waiting for five years to have a conversation with this lady," Margaret said. Whatever Garcia's personal history, Margaret saw someone from a place like West Oakland, who, with AB 617, had taken the methodology that WOEIP had been building for twenty years and made it into a piece of state legislation. "Community-driven, community-led citizen science," Margaret said. "Those things have never been related in any agency in California."

AB 617 was not going to be perfect, Garcia acknowledged. There were too many forces arrayed against the kinds of legislation EJ advocates really wanted. But, she said, "I wanted to start a way to harness the power that already existed in communities like this one and magnify it."

Most places hadn't been able to build something like this from the parts available in the legislation and their communities. But WOEIP did, and for Garcia, it was beautiful to see someone actually take advantage of the opportunity her legislation had created. "Everybody knows Ms. Margaret," she said. "Thank you. You are my idol. I so appreciate you."

The speeches ended. Brian handed out certificates to their most dedicated steering committee members. They were a cross section of the new Oakland that the Pacific Circuit has created. Tram Nguyen and Adriana Alvarado from Alameda County Public Health. The design firm Hyphae Design Labs' Brent Bucknum, the guy who introduced me to Margaret way back when. The pastor and local resident Reverend Ambrose Carroll. The local Latina politico Laura Arreola from the Port of Oakland. Richard Grow, the mustached EPA guy I met at Margaret's seventieth birthday. Fern Uennatornwaranggoon from the Environmental Defense Fund. The trucking world's Bill Aboudi, of course. And Alvirdia Owens, a resident who shows up.

People cracked cheap bottles of wine and we all ate these delicious Middle Eastern wraps, dripping turmeric-laced rice down our faces. Most everyone stuck around. It was a big old party raging over air quality improvement. Margaret bopped about, introducing me to different folks.

My own role in the Bay Area has changed over this past decade. I started hosting *Forum*, the big current affairs show on KQED, the NPR affiliate in San Francisco, taking over the nine o'clock hour from a legendary figure. Day after day, I talk with people who are struggling with the Bay Area's hardest problems: homelessness, housing affordability, institutional racism, climate change. And in that context, it can feel like there are no answers to the hard questions, no ideas worth pursuing, no policies that are succeeding. A label has arisen for the spiraling multifocal crisis of the San Francisco Bay Area: the Doom Loop. Like the blight of the urban renewal era or the "urban crisis" of the 1960s, the Doom Loop has become a way to talk about the brutal reality that there are problems in our communities that are completely outside their capacity to solve. Maybe it's all one long urban crisis, never solved, only hidden.

In the early days of the pandemic, I called my best friend and said, "I'm not sure how America is going to win against this virus." And he said: "Maybe you need to think about *how*, not whether, we're going to lose." Oof. Maybe that's the kind of gritty realism of Margaret and Brian that allowed them to win with AB 617, even if the legislation was set up for them to lose.

As I left the WOEIP meeting, I passed again by the two unhoused guys outside. One was still sitting in the car, but the other had set up a whole camp-stove cooking arrangement on the sidewalk. Steam rose from the pasta he was straining out of boiling water. "Yeah! Now it's time for the gravy," he said, dropping the pasta into tomato sauce. The smell and the scene stuck with me all the way to the car: the simple pleasure of cooking for someone and of blowing on a hot fork full of good food, anticipating that first bite.

Just a few blocks west of DeFremery Park, across Mandela Parkway, and nestled up against the old train station, the largest encampment in Northern California grew on Caltrans and city land. Wood Street, people called it, after the street it bordered. There were hundreds of people gathered into the informal settlement, pushed and pulled there by various forces and city authorities. Sometimes I found myself down there, surrounded by not just a few unhoused people but an entire outsider society stitched along the edge of the city, and just north of a shiny new housing development.

One of my colleagues at KQED, Erin Baldassari, spent months getting to know the people down there. After a fire broke out, the authorities—with funds from Governor Gavin Newsom—dismantled the bulk of what was there. After a tremendous struggle and much legal wrangling, the city offered the Wood Street folks what they call cabins, and most of them took the deal.

I can't say that I disagreed with the city's position. I still remember seeing my first homeless encampment, in San Francisco's SOMA district in 2007, and being horribly disturbed by it. It was so dystopian and unbelievable. Now my kids take encampments to be a standard feature of urban life. But I was also torn up by the stories from residents.

With Baldassari's help, our team put together a show that we broadcast from the remnants of the Wood Street encampment. When I arrived, there were small burning trash piles, stray dogs wandering around, the lingering smell of rotting food. Flies hovered over an arrangement of dilapidated couches. My lead producer got into a dispute with the lawyer for a group of Wood Street residents over the structure of the show, and thirty seconds before we began broadcasting from a folding table, the bulk of our guests walked out on us and wandered off down the street.

On air, we talked with Baldassari and a local homeless services provider for a while, and then, with twenty minutes left in the show, the Wood Street residents came back and sat down. They were quite a crew: multiracial, multigenerational. They'd been fighting to stay at Wood Street for years by that point.

The first person from their group to speak was a small white woman named Freeway. She and her husband had been living down there for *five years*.

I asked her what worked for her at Wood Street. "Everything," she said. "Everything worked for us. The whole time we were at Wood Street, we never had anything stolen from us. We were never in any real danger. We never even locked our door. Since we've been displaced, our van has been broken into five times, cumulatively."

I asked, what did that feeling of safety do for you? As Freeway answered, shakily, she held hands with a small woman, Minaz, and a large man with locs, John Janosko. When I glanced over, I saw they were all squeezing *so hard*, and he was crying silent tears. "It gave us the grounds to build our lives back," Freeway said. "The thing about Wood Street. It's not just a community. It's a family. People who haven't seen it can't understand, but this is the first place that I've felt like I had belonging and I had purpose."

It was hard to watch and hard to hear: the simple desire for belonging paired with the postapocalyptic setting where she'd been able to find it. There'd been dozens of fires at Wood Street in previous years. There was no running water, no heat. They were on industrially degraded land in life-eroding conditions. What killed me was that their solidarity—and the messy realities of improvised settlements—

contributed to the sense for most housed people that society was falling apart.

And yet: for these people right in front of me, holding one another, this was the best they'd had it in years. "We're not Disney villains. We're not the creepy guy in the trench coat that is gonna sneak up on you," Freeway said. "We're human beings. A lot of us have been through a lot of trauma and we are just trying to build our lives back."

Minaz spoke next. She was older. Maybe fifty? It was hard to tell. She was tiny and dark-haired. She radiated calm and elegance. And also, her hands and fingernails were dirty and her clothes were a step up from rags. I asked what the city could do to bring some of their sense of community into the city's official homeless services. She said they'd asked for a badminton net and a place where they could sit and talk and ask one another, "Did you make this a good day?" They wanted to build a real community that had a purpose. "We'd live every morning and ring that bell and say, 'Yes, here's what I'm here to do today,'" she told me. Instead, they didn't even have toilet paper sometimes.

A listener tweeted to criticize offering people like them housing. "I have to say that I disagree with the tweeter," Minaz said. "Having that place where someone can unclench their fists and look at their hands and say, 'I can make something with these hands again. I can rebuild something within myself.'"

You don't forget something like that. Who does not want to unclench their fists?

The deeper you get in, the harder the problems seem.

The whole system of the city is broken, isn't it? But then you see how hard people are trying. You have to squint a bit, but the community and purpose of the AB 617 crew, their desire to do right by West Oakland—was it so different from what was going on at Wood Street?

There will always be new challenges. Researchers coined a new phrase for what is going to happen in places like West Oakland long before sea level rise floods anything: toxic tides. As the mass of water increases along the shoreline, it will send groundwater pushing up, re-mobilizing toxins from past eras that might have otherwise lain dormant. The ghost pollution will rise. Hundreds of sites around the Bay Area could be affected, including dozens in West Oakland. Where are we gonna get the resources to clean all that up? The Doom Loop answer is obvious: we just won't.

But it feels to me that there is, right along the edge of the port, which is to say the shoreline, which is to say the edge of disastrous climate scenarios, the possibility of the kind of grand bargain that could lessen historical wrongs, protect our communities' health, and sustain the economies of places that had been sacrificed for economic growth.

This solution is deeply linked to the land, and I'll outline it in the conclusion to this book. The old working waterfront is facing a long crisis: hundreds of sites need workers and money to get cleaned up. Why couldn't newly remediated land get placed into a permanent real estate cooperative? New residents could pool their money with old residents and create stable, integrated, cohesive, clean communities with no market expiration date.

When it comes to climate, Margaret wants to see justice embedded in the movement as part of a cohesive vision for how our societies should work in the middle future. "Show me the whole picture!" she said. "Not this patchwork. We're living from a patchwork and emergency process. And that's part of the problem."

I wanted to talk with Phil Tagami one last time. Even in late 2023, the coal export terminal had not begun construction. His company had won ten legal battles in a row versus the city in different venues, but they were somewhat empty. They'd asked for more than $150 million in lost profits from the project, and gotten not even a million in the judgment.

He was moving back toward the edges of local politics, though. He went to a big San Francisco fundraiser with Mayor Thao, and the very night we met at the Rotunda building, the city of Oakland held a party there. His company had been up to other things, aside from just the massive terminal.

The reason I wanted to see Phil was simple: throughout all these years, I'd had a really simple question that I'd never gotten answered. Why didn't he just cut a deal with the city? At any point along the almost ten years of brawling, he and the city could have simply renegotiated the lease. Tagami would agree that his tenants would not transport coal and the city would change the financial arrangements of the lease. I wanted that *for the city and the climate*. And I wanted that *for him*. An Oakland without coal but with Phil Tagami is a more interesting place.

We made a plan to meet at the Rotunda again. I parked along a side street and said a small prayer that my windows would be there when I returned. Almost no one was out on the streets, even in front of City Hall. Was this just how things were and always would be?

He met me in the lobby. I was delighted to see him moving much better than in our previous interviews. Then we headed down a floor to a place I'd only heard about, a semi-secret wine cellar that he maintained in the basement of the building. Off a nondescript, fluorescent-lit maintenance hallway, he opened a door into . . . a miniature bar? A man cave? On the right side, he had a legit wine cellar and fully stocked bar, a bottle of Clase Azul standing tall on the bottom shelf.

On the left side, a quartet of comfy chairs were arranged around a coffee table. On one wall, there were two tacked-up maps of French wine regions (I remember only Côte de Nuits), which Phil told me he was trying to master. "Ignorance is curable," he said. On the other, a wall of framed rugby jerseys hung. On the floor next to my chair leaned an old Jerry Brown presidential campaign poster: VOTE FOR YOU, it read, delightfully weird white on brown.

We settled into the chairs with a nice burgundy and began to talk. I'd come very close to the end of this book process, which meant I'd reviewed thousands of pages of his documents, watched him testify in court, consorted with his enemies. I knew basically every public thing about Phil and many private ones, but people are vast cargo ships of stored knowledge and experience. And he still surprised the hell out of me.

It was in that meeting that I found out he was friends with Chappell Hayes, the environmental justice leader, a fact that I was quickly able to confirm. Tagami told me *he* was the guy who made the decision to name the observation tower at the port I'd lingered on so many times after Hayes. I asked him what he thought Chappell would have said about his bulk terminal. "He would have been upset," Tagami admitted. "And we would have certainly had a lot of conversations over everything that's unfolded at the Oakland Army Base, but he would have been coherent and thoughtful and it would have been constructive."

He schooled me on some older Oakland political figures, like Frank Ogawa, after whom the plaza outside was named—"a nurseryman, first Japanese American on the city council post–World War Two." We talked about the environmental conditions on the Army Base. I asked what the team found out there while doing the cleanup. "The question is more like, 'What didn't we find there?'" Tagami said. "It's basically an Army debris field."

Finally, I asked him the question that had been killing me. Why didn't he cut a deal with the city?

The short answer is: they tried and tried and tried. At several points, Tagami told me, they did seem to get close to a deal, but it never worked out. He blamed various people in the city apparatus. The city didn't want to take the loss, which would have probably meant some kind of monetary settlement with the Tagami team. And the longer the city fought, the more likely it would be that the different terminal lessees or Tagami himself would go bankrupt. The city could lose in court, but win the war of attrition.

Unlike in our previous meetings, Tagami seemed less angry, less defiant. He was funnier, a dark humor.

What was it like to have pulled yourself so far up the capitalist gradient of this world, and then end up like this? Just as he was about to grab the ring (or shit in the swinging jug, as it were), the whole city turned against him for being the one who wanted to build what the market wanted.

Couldn't he just stop? Couldn't he just give up? I asked. He said something like, I have to finish the job. That's *what he did*. That's *who he was*.

In the fight against climate change, not all the stranded assets will be coal mines. Phil Tagami has to learn everything for himself.

I've been to a lot of celebrations of Margaret now. A bunch of birthday parties. A gala for the West Oakland Health Center, where she got an award. Even the AB 617 celebration was also a celebration of Margaret. We'd danced and drunk, toasted and speechified. Brian once catalogued her catchphrases: "Closed mouths don't get fed." "If you're not on the table, you're on the menu." "Keep your head on a swivel because you never know where it's coming from."

At her seventy-third birthday, her son Zuri was there in a bright red-and-blue beanie, fresh Jordans, looking solid, despite all the time he spent driving for Uber and Lyft. He began his toast to his mother right as the family started organizing themselves to take pictures.

"Who knew my mom before West Oakland?" he asked. "Raise your hands."

As it turned out, nobody but the family gathered around had and they were more eager to get a picture than talk about those times. So he cut his speech short, handed over the microphone, and got into formation for the photos. When I asked him afterward about what he was going to say, he said he was going to talk about how she had battled their landlord in East Oakland over rats and other problems. He was going to say: she'd always had that fight in her. It's amazing how time can transform an Ingleside girl into a West Oakland woman.

Her legacy wasn't any one accomplishment, Margaret told me. No, it was in learning *and teaching* how to ask the right question. "Nobody seems to help you connect the dots and close the gaps. You got to go bottom up through the different levels and stages, to get an understanding of how to ask the right question." You wanna know how diesel trucks doing the work of the Pacific Circuit came to give kids asthma?

You gotta start by knowing Rudolf Diesel. "I went that far back. People don't know how to go back to come forward," she told me. "And when you stumble and fall, you go back again, and come forward again."

In the little archive of her papers she gave me, the only thing that she'd kept from her life of activism was a single business card she had from her time as a port commissioner. Because *that* was her greatest accomplishment. "I'm a Black woman off Seventh and Willow, right dead heart of the hood. And I was in this place of decision-making," she said. "Not just giving feedback and input but making a decision. That's a difference." Getting pushed off the board hurt a lot. Getting invited to the White House twice helped salve that pain.

You know, though, when Ms. Margaret passes on from this earth, which will happen, though it seems impossible, I don't want to remember other people gilding her name. I don't want to remember her fighting and/or working with the institutions of Oakland. There will be papers and news stories for all that.

No, I want to remember her turning seventy-two, celebrating with a big party at Seventh West, the newish Filipino-owned club along Seventh Street, the first real one in decades, its exterior walls painted with murals of containers and trucks.

It was awkward at times, competing versions of white "Happy Birthday" and the much better Stevie Wonder Black "Happy Birthday," but Margaret was unbothered. She was in her element.

When the required activities had been completed, some mid-tempo '80s R&B came on and Ms. Margaret shuffled to the dance floor. Her knees were going—she had surgery on them shortly thereafter—so she was planted firmly on one patch of floor.

Her hips could still move, though, and she swayed back and forth perfectly in time, happy and alive.

26

Conclusions

The city has been designed according to geometric, palliative,
symbolic and geographic principles, but increasingly urbanization is
organized through flows—flows of people, information, goods and
services across multiple scales of territory.
—CLARE LYSTER, *LEARNING FROM LOGISTICS*

The Pacific Circuit has crossed a threshold. It is breaking our cities.
"The city is based on exchange. It is a place that allows you to meet and
gather. That's the origin of the city," the historian Martha Poon told
me. "The commercial model of life is gone." How did we get there? It
was, in large part, "the application of these wartime logistical sciences
to the management of cities," she said.

Most people don't immediately connect container ships to rising
rents, jarring violence, and the attenuation of the city. But the connection is real and strong. The logic of the system makes some choices
easier and others harder, reshaping the layers of our metropolitan areas. Other trends are at work: phones are everywhere, the untended
wounds of institutional racism fester in the deeds of our cities, political polarization makes legislative redress of anything fantastically
difficult.

But all those causes of our modern era of urban crisis have received
enormous attention. The economic structural changes that logistics,
broadly, and transpacific supply chains, specifically, have wrought in

our cities are wildly underrated. Here, in Oakland, the birthplace of the whole thing, life has become grueling for most, despite some of the highest incomes in the country.

The same ideas about how to cost-optimize bombing runs in Europe and manage supply shipments to Vietnam were applied to the scoring systems underpinning urban renewal and to the large-scale production and distribution of a vast array of products. Then, filtered through phones and data centers, the same tools came for the very idea of restaurants and the other low-key semipublic spaces that make a city *a city*. Internet entrepreneurs flaked off the profitable parts of urban businesses and cast off the weight of the positive externalities they generated.

Coastal cities had seemed resilient. They were growing and increasingly rich. Our dense patchwork of businesses didn't get knocked out by the big-box era of the circuit's functioning, and money flowed steadily into the narrow sliver of real estate along the Pacific.

But we are in a new time after the pandemic. The logic of logistics has continued to extend down to the household level. Everybody with a phone is part of these networks for moving people and goods. Amazon allows a Chinese manufacturer to direct something right to your house, bypassing any local business not involved in the movement of goods. Uber (and the Uber-for-everything apps) takes you anywhere or brings anything to you. It's harder to open a shop selling just about anything. It's harder to run a restaurant. It's harder to keep public transit running when it's bleeding riders to apps.

But the Pacific Circuit hasn't only changed cities. It has also changed us. There is such a thing as the logistical self now. These logistical figures, as the architectural researcher Clare Lyster calls them (us?), are people who understand themselves as nodes in a network of flows, less citizens of Oakland or San Jose than delivery or pickup points across a variety of apps. One does not wander the city like a flaneur but draws on the whole world's resources to deliver whatever is desired. People who lived in big cities used to love "having a guy" who had access to niche products or services. Now there is always an app, a website, and then a delivery. Instead of a network of trusted people across a metropolitan region, there is . . . a data trail and a recycling bin filled with cardboard boxes.

It's an economic change, but also a social and a psychological one. Just ask Kurt Vonnegut. There's a story he tells that goes viral every once in a while. It's about a time when he goes out to buy an envelope and his wife asks why he doesn't buy a hundred envelopes and keep them in his office. (That optimizing, logistical thinking!) Vonnegut has no intention of doing so. "I'm going to have a hell of a good time in the process of buying one envelope. I meet a lot of people. And see some great-looking babes. And a fire engine goes by. And I give them the thumbs up. And ask a woman what kind of dog that is," Vonnegut told a PBS interviewer. "And, of course, the computers will do us out of that. And, what the computer people don't realize, or they don't care, is we're dancing animals. You know, we love to move around. And we're not supposed to dance at all anymore."

That's it, basically. Cities are so good at building solidarity: not the hard solidarity of getting into the streets in the interest of others but the soft one of building human allegiance for no reason at all. With an exchange of money, there are also the inefficient interchanges of life. Longshoremen were famous for their bullshitting, the verbal dancing of talking animals. Cabbies and barbers, too. But city life is filled with all kinds of little interactions: grocery checkers, the bookstore people, a Japanese steel specialist who carries obscure gardening tools, the vintage store lady with an eye for French chore coats, the guy selling *Street Spirit* outside the cafe, a barista with an eyebrow ring and a heart of gold. That woman who says, smiling, "Why, this dog? He's a chiweenie."

This economic model I've laid out, the Pacific Circuit, born out of wartime need in Vietnam, controlled by Silicon Valley technology, and spreading out from its American birthplace, Oakland, really has rewired the world. I want to lay out the logic of the system here because then, I hope, you will see it everywhere, as I do now.

First and foremost, it drives relentlessly toward scale and flow and technologization. Amazon is predicated on being the biggest and the fastest. Hence all the warehouses built by Prologis (now almost a $100 billion company in its own right). Apple must turn over its merchandise nearly as fast as the company can manufacture it. Ships *have* to keep moving. Every day of delay is hugely expensive. Normal human timekeeping does not apply, whether or not sailors would like a couple more days onshore before setting out back across the Pacific. Note that it is not speed per se but the consistent, synchronized flow that allows global supply chains to work. Things arrive right when the factory or the store needs them.

As part of that need, the Pacific Circuit seeks to commodify everything, packaging it in ways that obscure the differences between items in circulation. The more everything is the same—just a packet of something—the easier it is to build systems that can model those components and move them around. This tends to make these systems opaque: one Uber is all Ubers, one banana is all bananas, one container is all containers, one warehouse is all warehouses. The hard reality of our world is that real life frustrates these abstractions, but that doesn't mean logistics companies, very broadly construed, can't try to work the kinks out.

It is completely insensitive to ethical or even national imperatives. As Huey Newton observed in his essay "The Technology Question," the moral responsibility for global supply chains becomes so diffuse as to be unrecognizable. And we can love the outputs of a system that we hate. There's no way to escape living within the Pacific Circuit. Social justice organizers shop at Target, too. But most of the time, observers of globalization have concluded that it has further advantaged rich countries. Even if workers in the United States became less powerful and had to take lower wages, Americans benefited from cheaper products produced in Asia. To some extent, this is definitely true. Have you seen the price of a huge television recently? But the system has spun far beyond the control of the United States, which might have begun its creation. It is not a simple top-down hierarchical system. It works through complex feedback loops and network changes. To the nation, the pandemic showed that it's not good to be entirely dependent on other countries for producing essential goods. Perhaps it isn't good to outsource many types of products. The Biden administration's new rules on electric vehicles, say, or semiconductors are a form of protectionism (or industrial policy) that would have been unthinkable during the Obama years. But it's not so easy to unlink from the Pacific Circuit. There are so many connections among the nations and companies and people involved that it will take many years to rebuild a different kind of domestic economy, if that's something the national government actually undertakes.

Most important for our story, the Pacific Circuit relentlessly externalizes environmental costs into environmental sacrifice zones both in Asia and the United States. That has generated enormous problems for the health of people across the world, and specifically in West Oakland. Their bodies have been knowingly polluted in exchange for economic growth that benefits a small subset of people. That is the deal, and it has to be recognized as such.

Go wide, geopolitical and planetary wide, and you see a different set of issues coming to a breaking point. The coal that might eventually run through Phil Tagami's terminal? It will go to power export-oriented economies that will use it to make the goods that will get shipped right back through Oakland. That lets us offshore the emissions and pollution that our consumer economy generates, just as semiconductor companies were able to offshore the ugly business of assembling huge workforces to do difficult handwork for decades. It's out of sight, across an ocean. But we share just this one world, and greenhouse gas emissions have reached levels almost certain to cause dramatic and negative effects for humans. Worse, the naive and ahistorical approach that Americans have taken with respect to climate negotiations counts on blaming China and other developing nations for burning fossil fuels, even if a big chunk of that fossil fuel ultimately ends up embodied in goods flowing back across the ocean to feed U.S. needs.

The Pacific Circuit makes power liquid. It is a system where anything can happen, but nothing can be done.

←---   ---→

Not everything that has come out of the Pacific Circuit is bad. Most significantly, the creation of this ocean-spanning network of networks has placed people in different countries and cultures into close proximity. Workers might even be on the same assembly line, as it used to be put. There are vast links between different Asian communities and specific places inside the United States. Huey Newton's last great political realization was that the global landscape of power was no longer only one of nations but an interlocking set of "intercommunal" interests. The conditions of one neighborhood in Oakland might be shockingly similar to one in Gaza or Rio de Janeiro or Seoul. We might be separated by our identities, but we can be united by place and our actions within it. To care for the land and all its histories and living beings is to begin an intercommunal relationship that can extend all across the world. And that's the basis on which resistance to the Pacific Circuit and the conditions of liquid modernity can rest.

The land is the key. The land itself cannot be rearranged for the benefit of logistical flow. It is also elementally different, parceled and propertied out. Power sticks there. People get attached, too, and they're willing to fight for the places where they live. I don't mean blood and soil, which is dangerous, but simply to care for the places where we live, regardless of the orientation to it you were raised with. For me, this is the potential of environmental justice. It can't be seen as just a movement of particular hazards in particular places. The potential of the idea is so much greater. These production systems are always pushing power and decision-making away from the very places that have to endure the problems they create. Environmental justice inverts that, and says that those who are on the land, in the place, taking care of the neighborhood, *those* are the people who must be the crucial decision-makers.

Logistics is, fundamentally, a system of flows, an attempt to deterritorialize, to let economies push through whatever they want. To make the land work within the Pacific Circuit, it has to be stripped of its specificity, turned into a container of economic objects. That's what Prologis does, after all. The shiny new warehouses right there on the Army Base and all over the world might as well be interchangeable. They have turned pieces of land into assets. In a slightly different way, that's what mortgage-backed securities do, too. They roll up all these homes, these lives, the specific conditions of millions of years ago and thousands of years ago and sixty years ago into . . . an entry in a data-base representing a cash *flow*.

Like the body, the land keeps the score. That's why caring about a cafe and caring about the destruction of communities in Malaysia and caring about a feral lot filled with more-than-human life and carcino-genic chemicals are all of a piece.

What has been done *to a place*, the way it was blighted, that's all in the ground and circulating in the air. Reparations are normally pre-sented as a way of delivering resources to oppressed people—in the United States, African Americans—but if we look back at U.S. policies around property in the twentieth century, they actually targeted *the places Black people lived*, including anyone else who lived nearby. With that knowledge, we might say that reparations could be place-based. The land could be repaired. If we look at the provisions that were attached to the Oakland Army Base redevelopment, we have something of a model for how to build real economic empowerment into this work. We'd make sure that there were local hires, that there were lots of local busi-nesses involved, that citizens returning from incarceration had a fair shot at getting jobs and training. There is so much work to be done, and as Ms. Margaret's organization, the West Oakland Environmental Indicators Project, has been arguing—that work can become careers for local residents.

Imagine this remediation as a route to reparations, and when that land had been returned to health, it would come into the owner-ship of a cooperative of long-term residents, who had been subjected to the conditions that placed toxic chemicals on the land in the first place. Finally, they would not be part of an advisory committee but an ownership group. Around West Oakland, that would largely, but not exclusively, benefit Black people, just as the policies of the twentieth century largely, but not exclusively, targeted Black people. That's the model of reparations that I think has the greatest chance to succeed in a multiracial democracy.

We could even look at the mid-century blight calculations that I've argued were an index of environmental racism, and concentrate these reparative efforts in precisely the areas with the highest "penalty scores." The receipts are all there, in the archives, and in the chemicals that have leached into the ground.

What happens from there would be up to the people who have lived in that place. We have fascinating examples of different groups in the East Bay who have a different relationship to land and property than the simple market mechanisms that have dominated efforts.

There has been a growing movement among the indigenous people of the East Bay to "rematriate" land, as they call it. Drawing on a voluntary land tax they call "shuumi" as well as major contributions from the Kataly Foundation (which is doing all kinds of interesting work in this area), they have managed to acquire urban land. So far, they've dedicated much of it to ritual purposes, but who is to say they will stop there? The Squamish Nation in Vancouver, British Columbia, is building six thousand housing units in huge towers, a development that's possible because of their independence from land-use regulations.

Land trusts are popping up all over the Bay Area, from the far-out agricultural regions to the innerest inner city. But there's a new intriguing idea, incubated at an Oakland institution called the Sustainable Economies Law Center. One of its co-founders, Janelle Orsi, came up with a legal framework for answering the challenges posed by the history and present of the financialization of property. She called it the "Permanent Real Estate Cooperative." "Our vision is for a typical PREC to have hundreds or thousands of members who look around at the land and buildings in their community and think: 'We should own this!'" the SELC wrote. "Rather than watching the fate of their communities be determined by wealthy speculators, large companies, and absentee landlords, PREC members will build collective power, pools of capital, skills, and organized communities that can take action to shape the future of local land and buildings."

It's a thrilling vision for countering the forces that have been tearing apart our cities. Big cooperatives—not nonprofits—can collectively purchase homes and commercial property and move them out of the market through the legal framework established by Orsi. In theory, they could stabilize prices and provide more resident control, all within a democratic structure. It also makes the interwoven poverty and wealth of West Oakland an asset, instead of a corrosive force. Rich people can do more than drive up prices. They can invest in permanently affordable, cooperative housing.

And who is the executive director of the first permanent real estate cooperative, the East Bay Permanent Real Estate Cooperative? Noni Session, whom we met first as a child living in the shadow of the Cypress Structure, and then again, transforming the local politics of West Oakland. In September 2021, after building the resources to acquire several homes, it purchased the Esther's Orbit Room building.

This alone won't end the long urban crisis. To do that, a multiracial coalition has to be built to end residential segregation and narrow the Black-white wealth gap. Taking segregation first, Heather McGhee's *The Sum of Us* has convinced me that the reason we can't have nice things in this country is a racial hierarchy that relies on placing Black people outside the circle of civic, social, and human concern. The means for enforcing this terrible and destructive lie has been residential segregation. As Douglas Massey and Nancy Denton incontrovertibly showed in *American Apartheid*, no other racial or ethnic group has *ever* been as isolated as Black people have always been. While this has led to some measure of limited political power in some places, *separate* will be *unequal*.

The great tragedy for the country is that in positioning Black people in certain places to protect white wealth in others, we built a poison pill into social policy after social policy. The fear of Black people receiving more than they are perceived to deserve keeps everybody from sharing fairly in the incredible wealth that this country has generated. This problem cannot be solved individually. It cannot be solved by white people being nicer to Black people or by a cop hugging a child. This is a collective problem that requires restructuring our economic and social policies.

The sort of reparative policy I imagine would spiral outward and build the coalition supporting it. The longtime Oakland activist Angela Glover Blackwell has a beautiful model for this kind of politics: the *curb cut*. As told beautifully in the film *Crip Camp*, disability rights advocates working from a home base in Berkeley began to make the case that reshaping urban infrastructure to suit the needs of people in wheelchairs was both a moral good *and* something that would benefit many other people.

Here's the model: first and foremost, curb cuts served people who were tremendously marginalized and gave them new freedoms and life possibilities. Second, over the decades, disability rights advocates were able to actually change the physical infrastructure and plans of cities, making their wins durable. Third, they were able to enlist the power and scale of the federal government in encoding their needs across America, at considerable cost to private entities.

Yes, we need the state, and a state with far more capacity than it currently has.

Cat Brooks, the Anti Police-Terror Project organizer in Oakland, has a saying: "All violence is state violence." The lawyer and essayist Savala Nolan talks about how "the state and its extended, strong, pilfering fingers has always had a grip" on her family. There's a reason that the libertarians and the Black activists of West Oakland made common cause against urban renewal. State power can and often has been destructive. "The state, the state," Nolan writes, "maker of slaves, trafficker of bodies, thief of children, grave of memories, grave of families, wicked and bigoted decider of fates, reaching forward and backward in time."

Ms. Margaret's family's history does not contradict this analysis. What do we have otherwise, though? Reduced state capacity since the years of high liberalism has not meant more power for the disempowered but increasing inequality as the market crushes people.

The Pacific Circuit has made it far harder to get a handle on power. The real fight is not with the mayor or even the Port of Oakland.

There's a bigger set of actors. Even a union that controls a whole coast's worth of ports cannot exercise power in the way it might have. The kind of power that structures much of American life now has no political form. You can't vote against it. Boycotting it would be nearly impossible. The first task, perhaps, is to make this system of power visible and knowable.

Then the deep-rooted work of battling state power to *direct it* along the pathways that communities actually need can begin. AB 617 had crucial limitations *and* it could be retooled to do good work. Rooting out institutional racism is necessary precisely because then the necessary scale of state power can be fairly deployed.

Reparations are a necessary first step into our collective future. Every part of our lives has greenhouse gases embedded in it, as it has the muscle power of enslaved Africans, the blood of indigenous people, the toil of underpaid women, and the increase in white property values created by segregation. We learn to pay homage to those ghosts, or we'll be haunted forever.

This book has tried to pull together the ways that people living in and through the system have recognized, worked with, and fought the power embedded in this transnational system. Maybe it has made a smaller argument, too, that even within what we call Capital-C Capitalism, there are versions of the way it works that may be more or less compatible with the communal life of the city. The capitalism that began building in the '60s, accelerated through the long tech boom, and became ubiquitous during the pandemic is tearing apart the very basics of urban life.

During the nine years I worked on this book, Oakland's fortunes have risen and fallen. Our city, once again, is a poster child for the urban crisis, and it is also one of the most beautiful, prosperous places in the world. If the Oakland of 2025 proves anything, it is that the traditional model of urban development no longer can result in prosperous, safe, and equity-enhancing cities. Oakland's median income has skyrocketed since the 1990s. Our housing prices have jumped even more. There's *so* much more money flowing through the city than there was in decades past, and yet the murder rate is comparable to that of the late '90s. Property crime is extremely high, especially car and retail break-ins. No one knows precisely why crime goes up and down, but there's no doubt that Oakland's crime problems have been unexpectedly bad. If the wholesale replacement of residents through gentrification doesn't make a city safer, what does that say about that model of development?

Perhaps the towers and corporate landlords and redevelopment out at the port—the big shit, as Brian Beveridge would put it—did almost nothing for the poor people of the city, or at least not enough to offset the rising cost of living and the effects of the Pacific Circuit. You can pave over the history, but it is still down there.

The long "urban crisis," initiated when Black and then other migrants arrived in American cities in the mid-twentieth century, has never really ended. All the solutions of the past hundred years were shortsighted and worked for one generation and one race only. If we're going to build a material world for supporting multiracial democracy, the Pacific Circuit might be the only structure up to the task. Perhaps it can be rewired to serve people's basic needs or people can change their relationship to its logic.

I think I know the values that would need to be built into this system of efficiency and flow. I observed them among people who have lost everything and are living in the homeless encampments that are the symbol of our times. Amid all the deprivation and difficulty, they are strengthened by two things: community and purpose. Let us begin there, then, as we imagine new, collective ways of making our cities work.

Epilogue

Let me take you back out to the port. Let's watch the cranes and birds before we go see Noni Session at Esther's Orbit Room. We've been here before. We took this trip at the beginning of this book, looping down to the park in the middle of the Port of Oakland, just a few hundred yards from the Chappell Hayes Observation Tower, right up against the action of the port.

There's always more to see.

To our left, a giant white crane is unloading a ship (it could be any ship). A tiny human directs a tiny glass-bottomed room back and forth along the long boom of a crane. The operator stops over a container and drops the spreader with a joystick, cables reeling, as it goes down, down into the hold. It locks on to the corners of the box below and the crane hoists it out of the belly and into the air.

You know this place birthed a new global mode of being. Containers, computing, and the sacrificial landscape just over on the other side of the freeway. The Seventh Street Terminal, the last land to be reclaimed from the bay, and the first home of Japanese shipping companies in the United States, is right there.

In the distance, San Francisco glitters, misted in an elven haze. What a place. Run your eyes down the peninsula. So much of the technology that was created there has been lashed to the Pacific Circuit and used to immiserate and to liberate.

Sometimes I can't believe I've been coming out here for a third of my adult life. Why do I keep coming back? These vistas. I really do love this place. These machines, glimpsed from the Bay Bridge or hiking in the hills, can seem like exotic birds that paused here during some global migration. Up close, the physicality of them is startling. The smell, the verticality, the sound of steel on steel, low, deep oceanic vibrations.

Tugboats in the twilight, straining against their ropes to slow a vast beast in the shipping channel.

A dockworker, alone on a break, silhouetted at the end of a jetty, tiny ember in his hand, looking out over Creation.

A vast line of trucks filled with the living people of earth, each a story pulsing through the port.

When a ship passes, there is a great emptiness in the channel, then the water returns, as if nothing was ever there.

The port is a wasteland, and it is also beautiful. It's not made for humans. People are capable of creating things that have their own needs, which are controlled from elsewhere, barely responsive to the tiny people looking up. And yet here we live our lives, dreaming of horses, turning over a conversation, trying to forget, being in a body, wanting to quit, growing up, growing old, remembering clouds.

I stare at the sides of these ships coming to the dock, at the layers of paint chipped, scratched, pockmarked, hints of rust, and all I can think is: this is a planetary memory.

Let's head up Seventh Street, under the rebuilt highway that used to be the Cypress Structure, past that hulking post office. It sounds strange to say it out loud, but I don't find it hard to identify the three key problems of our time: closing the racial health gap, decarbonizing our world, and reining in runaway inequality. On the ground in Oakland, working on any of them requires working on all of them. These are imperatives for the survival of a healthy society, a living planet, and an America that can regain a sense of future possibility.

I can't stand there on Seventh and say African Americans don't have a claim on our country for reparation for old *and* new traumas. We did the math on how this place was racially targeted, and that's just one of *so many* situations like that. Encased in concrete or not, those histories live in the bodies of people and in the earth itself.

But here's the thing: reparations aren't only about reckoning with the past. They are about allowing our country to move into the future. We need a more equal society, so that we can mend its fabric, care for it, patch the holes. If we really want equality in America, reparations cannot be a onetime payment and an apology, which is what Japanese Americans received for their internment.

It simply will not be that easy to repair what has been broken. If *that* is true, we need a durable plan to begin reparation and continue it for many years. And if *that* is true—given the demographics of the country—we need a multiracial coalition to understand that it is in our collective self-interest to attack the problems that anti-Blackness created. The guilt of white people is not enough. What's required for reparation is *enormous*, and that means making the case that reparations are something we *all* need, despite our own ancestors' troubled relationships with the United States.

Ah, good, we've arrived back at Esther's Orbit Room. The sign is so old. Noni is here already in a fresh new black, red, and green jacket, braids pulled into a top bun. She's gotta do her regular trash pickup out front, and she has a pair of gloves for me, too.

But first, let's peek inside. I've been waiting for this moment for almost a decade, to go inside the last remnant of *that Seventh Street,* and see what Esther was up to in here. I step forward, eyes adjusting.

Everything is pretty much as it was when they closed the doors, which is how it was for the decades before that. There's a kitchen in the back, and some low-slung seating around the edge of the room. Green carpet, not nice. The ceiling is like the night sky—painted black with the suggestion of stars.

Over the years, things have stayed where they've been dropped. A fan. A folding chair. A bucket. A cooler. The layer of dust and disarray is somewhere between abandoned house and mummified post-COVID office.

But the bar is stunning. Mirrored and gorgeous, I can see Noni and me there together. There's an old cash register, predigital. A landline phone. Tangles of cords run up into the ceiling. Plastic flowers stand in a fake silver vase. The drink prices on the signs tacked onto the wood are hand-lettered. And most incredibly, the bottles that were there when Esther's shut down are *still there*. Every single bottle has had a piece of plastic delicately placed over its pour top and wrapped with a rubber band.

"That's hope right there," Noni says. "They knew somebody would come back."

A Long Note on Sources

This book began with a lot of wandering around Oakland, observing the dynamics of a city undergoing tremendous change. I talk with everybody, much to my children's chagrin, and I first became interested in writing about this city because it seemed to embody the kinds of urban change that were roiling the country in the 2010s. That interest became a podcast, *Containers*, for which I recorded dozens of interviews with different people on and near the waterfront. These are not cited separately from the work I did on the book, as I always envisioned it as one project.

My core informant has always been Ms. Margaret, and we had dozens of conversations over the years, big and small. She was also generous enough to share her personal archive with me, as noted in the book. I spoke with several members of her family, too. Though few are directly cited here, they helped inform my understanding of her life and times.

I stumbled upon a trio of older longshoremen—Brian Nelson, Frank Silva, and Mike Vawter—who each maintained a historical record of his time on the waterfront in the twentieth century. Brian Nelson had a series of audiotapes, which we digitized together in his basement. As he explained the different sounds that we were hearing, I learned about the processes of dockwork. Frank Silva had made gorgeous photographs of men at work, and these also helped me to grasp the waterfront life. Mike Vawter connected up the old college radicals and the life of ILWU Local 10, and he shared many stories with me along with too many margaritas at Little Star in Albany. All these were invaluable discussions in the production of this book, though they remain largely under the surface of the text.

There are several great books on the longshore workers of the ILWU. Peter Cole's *Dockworker Power* is probably the most germane to this book, though I also loved Harvey Schwartz's oral histories and Margaret Levi and John S. Ahlquist's *In the Interest of Others*.

Oakland historiography is spotty. There's a good one-volume history, Mitchell Schwarzer's *Hella Town*, which came out as I was working on this book. It's a great introduction to the city's past. Robert O. Self's powerful *American Babylon* excavates a great deal on the Black Panthers, Black Power, urban renewal, and the suburban property tax revolt in Oakland. It was so influential for me in framing this book that no number of citations could suffice. Chris Rhomberg's history *No There There* was also very useful for understanding mid-century Oakland politics.

I built a small archive of Oakland city planning documents from the collections at the Environmental Design Library at UC Berkeley, the Oakland History Center, and online collections. It's available on *The Pacific Circuit* website. Some reports are cited directly and others I used to inform my broader understanding of the civic and political climate of the city.

In order to get a more intimate handle on Oakland's history, I consulted the essential archives of the African American Museum and Library at Oakland (AAMLO). Along with works directly cited here, I also paged through or listened to many oral histories and other primary documents contained in the collection there. *The Combination*, something like a Black alt weekly, was particularly interesting in its depiction of everyday Black life in the East Bay. I mapped its advertisers to get a sense of the commercial corridors of the community. That map is available at *The Pacific Circuit* website. *Flatlands* magazine was crucial for describing the conditions that poor people faced in West Oakland. The entire run is available at AAMLO, and I soaked in its descriptions far more than I have cited it in this book.

Gretchen Lemke-Santangelo's work on African American women migrants to the East Bay, *Abiding Courage*, is indispensable, as are

Dorothea Lange's photographs of migrants to Richmond, many held at the Oakland Museum of California. To deepen my understanding of Ms. Margaret's family's experience, I consulted oral histories of Black migrants held at Berkeley's Bancroft Library, as well as personal histories held in the archives at the Rosie the Riveter National Historical Park. Some are handwritten. Others are typed professionally. A few contain photographs. Thank you to Priscilla Shelton, Margaret Archie, Mary Billups, Ernestine Joyce Shepherd, Elson Nash, Elaine Lolos Lackey, and especially Lurena-Ash Blair for sharing their wartime experiences. I augmented the personal histories with Richmond city reports as well as Katherine Archibald's remarkable anthropological study of race in the shipyards, *Wartime Shipyard*.

I also made extensive use of both Oakland and San Francisco public libraries' history rooms. In Oakland, the legendary Dorothy Lazard helped guide me to piles of clippings and individual documents related to the port as well as the Prescott neighborhood. Many of the sources cited here were discovered there. In San Francisco, my mother and I created an index for the run of *The Spokesman* newspaper, which covered Bayview Hunters Point, during the late '60s and early '70s. It was an essential component of understanding Ms. Margaret's milieu, the way War on Poverty programs ran, and the radicalization of those times. The index is available on my website. In San Francisco, I made extensive use of Wally Mersereau's collection of documents about BART's real estate operations. In addition to the things I directly cited, the many reports, brochures, and advertisements collected there paint an unparalleled portrait of the civic and bureaucratic project of building BART. The administrative offices of BART also generously allowed me time in their collections to look for more information about Ms. Margaret's neighborhood as well as West Oakland. I found and used a variety of documents and maps related to the taking of property in those areas.

Much Black Oakland history is embedded in books about the Black Panthers. Alongside Huey Newton's many works (and fascinating archive at Stanford), Elaine Brown's *A Taste of Power* and Bobby Seale's *Seize the Time* stand out among books written *by* Panthers. Ericka Huggins's 2007 oral history, held at the Bancroft Library, is clear-eyed and fascinating, too. There are so many books written about the Panthers, but three really shaped the way I approached the group in this book. For the ideology of the Panthers during the period of their formation, I turned to *Black Against Empire*, by Joshua Bloom and Waldo E. Martin. It is sweeping, comprehensive, and less caught up in mythology than many other works. Donna Murch's *Living for the City* situates the Panthers inside the Great Migration; it's similarly rich and nuanced. Robyn C. Spencer's *The Revolution Has Come* fills in important gaps in the gendered history of the Panthers.

Many characters were touched by the War on Poverty. The aforementioned *Spokesman* covered the San Francisco programs closely and helped me understand their local day-to-day activities. The Black Panthers, in some crucial respects, grew out of the War on Poverty, quite literally from the offices at 55th and Market in North Oakland. I am grateful to archivists at the Oakland History Center, NARA regional archive in San Bruno, and California State Archives for their help in finding documents relating to Huey Newton's and Bobby Seale's participation in Oakland's local programs, as well as the broader texture of the times. I am deeply indebted to Martha Bailey and the many other participants at the War on Poverty Conference held in 2014 at the Center for Poverty and Inequality Research at UC Davis. Their work completely changed my position on the success of Lyndon Johnson's slate of initiatives, which informed the tone of this work.

There are several important historical resources specific to Seventh Street in West Oakland because of its importance in musical history. Cheryl Fabio's film *Evolutionary Blues* is a great resource for understanding the development of Seventh Street during its heyday in the midcentury. An equally fascinating film is *Long Train Running*, by Marlon Riggs and Peter Webster, which covers the later era of decline with grit and beauty. The various publications of and about Arhoolie Records also provide some invaluable information about life in West Oakland, including the album *Oakland Blues*.

Similarly, there are important resources specific to the Oakland Army Base and its shutdown. In addition to the Base Realignment and Closure commission national reports and local documents, the Bancroft Library also produced a series of oral histories that were essential to this project. They were collected into a 2010 volume published as *The Oakland Army Base: An Oral History*, edited by Martin Meeker. Phil Tagami also provided a vast number of documents related to the litigation between his company and the city of Oakland, which I consulted.

In order to grasp the material underpinnings of what we call gentrification, I reconstructed a full dataset of mortgage loans made to different demographic groups in West Oakland, thanks to the Home Mortgage Disclosure Act. This required joining older and newer HMDA data for the zip codes enclosed by the freeways and generally agreed to form West Oakland.

Of course, Oakland is part of the larger San Francisco metro area, and several books on other cities helped me understand how my themes worked within them. Shirley Ann Wilson Moore's *To Place Our Deeds* is a great academic history of Richmond's Black community. *The Postwar Struggle for Civil Rights*, by Paul T. Miller, is a fine history of mid-century activism in San Francisco. I also consulted various documents produced by organizations like the Council for Civic Unity and the papers of civil rights leaders like C. L. Dellums, held at Berkeley's Bancroft Library, which gave some texture to the fight for housing equality in the Bay Area.

Several documentary films produced about the era's racial battles by local journalists are also remarkable. The very best of them is Richard O. Moore's *Take This Hammer*, hosted by James Baldwin and centered in San Francisco. Another outstanding report is *Segregation: Western Style*, made by KRON-TV, which describes the fight over housing discrimination in the Bay Area. Both films, and a remarkable amount of other contemporary television reporting that informed this book, are available through San Francisco State's DIVA archive.

Because Oakland's history has been so deeply shaped by redlining and other government interventions into the real estate and mortgage markets, I spent a lot of time trying to understand the mid-century government interventions that shaped American cities. There is, unfortunately, so much to learn about what the government did. The works I drew most heavily on to understand the system were *American Apartheid*, by Douglas Massey and Nancy Denton, and *The Color of Law*, by Richard Rothstein, which focuses much attention on the San Francisco region. I read and was influenced by Jennifer S. Light's *The Nature of Cities*, David M. P. Freund's *Colored Property*, Jeffrey M. Hornstein's *A Nation of Realtors*, Kenneth T. Jackson's *Crabgrass Frontier*, Robert M. Fogelson's *Downtown*, and Marilynn S. Johnson's *The Second Gold Rush*. I remain deeply indebted to Ta-Nehisi Coates for the groundbreaking work he did in this vein while we were colleagues at *The Atlantic*. In my analysis, the concept of *blight* played a large role in Oakland, and I was heavily influenced by the legal scholar Wendell Pritchett's paper "The 'Public Menace' of Blight" and a dissertation by Brian Robick, "Blight: The Development of a Contested Concept." Matthew Desmond's work *Evicted* and Keeanga-Yamahtta Taylor's *Race for Profit* were both crucial inspiration for this work, too.

There are many fascinating works about logistics. I was influenced by Clare Lyster's *Learning From Logistics*, Jesse LeCavalier's *The Rule of Logistics*, and Deborah Cowen's *The Deadly Life of Logistics*. I am also deeply indebted to two other scholars. Charmaine Chua's research weighs heavily on this book. And at the intersection of logistics and financialization, the historian of capitalism Dan Bouk directed me to Martha Poon, who changed the course of this book in a single conversation. John McPhee's rich interest in the workers inside the logistics system as compiled in *Uncommon Carriers* has long been an inspiration.

For containerization, the classic is, of course, *The Box*, by Marc Levinson. I found many fascinating materials within military collections, especially Sea-Land films recorded in Vietnam at Cam Ranh Bay. Tim Hwang's interest in and framing of containerization also fed into this work.

Adam Tooze, in his books and in conversation with me, deeply informed this book's approach to the state and state capacity. I also read and was influenced by *Capital City*, by Samuel Stein; *Unmasking Administrative Evil*, by Guy B. Adams and Danny L. Balfour; *Seeing Like a State*, by James C. Scott; and *Don't Let It Get You Down*, by Savala Nolan.

I have cited most of my key influences as they relate to the technology industry, but the historians Leslie Berlin, Dan Bouk, Paul N.

Edwards, Margaret O'Mara, and Christophe Lécuyer deserve extra recognition for helping this book travel outside the conventional narratives of Silicon Valley.

Lenny Siegel at the Pacific Studies Center also opened up his archive of local Silicon Valley activism to me and answered many of my questions. The center's research into chipmakers' supply chains in the 1970s is unique and deeply underrated. He also introduced me to the online archive of the April Third Movement at Stanford, which, along with *The Stanford Daily*, helped me to confirm the stories that Mike Vawter told me over margaritas. San Francisco State's online archives were also essential for telling the story of the Third World Liberation Front strike there.

Notes

INTRODUCTION

3 *the Pacific is now*: Mansel Blackford, *Pathways to the Present* (Honolulu: University of Hawaii Press, 2007), 5.

3 *their market values*: Einar H. Dyvik, "Biggest Companies in the World by Market Cap 2023," Statista, August 30, 2023, https://www.statista.com/statistics/263264/top-companies-in-the-world-by-market-capitalization/.

4 *$200 billion*: Blackford, *Pathways to the Present*, 20.

5 *$500 billion of exports*: U.S. Census Bureau (2024), "Trade in Goods with Pacific Rim," https://www.census.gov/foreign-trade/balance/c0014.html#2005.

6 *more than seven times*: Matt Levin, "What Ever Happened to the California Dream?," CalMatters, 2019, https://projects.calmatters.org/2019/california-dream/.

6 *single largest racial group in the Bay Area*: Jay Barmann, "New Census Data Shows That Asians Have Overtaken Whites as Largest Bay Area Racial Group," *SFist*, May 25, 2023, https://sfist.com/2023/05/25/new-census-data-shows-that-asians-have-overtaken-whites-as-largest-bay-area-racial-group/.

6 *by 2020*: Metropolitan Transportation Commission, Bay Area Census, http://www.bayareacensus.ca.gov/bayarea70.htm.

1. MEET MS. MARGARET

11 *a short, nervous speech*: Obama White House, "Champions of Change: Citizen Science," streamed live on June 25, 2013, YouTube video, 1:26:43, https://www.youtube.com/watch?v=PLau1ZFA8z8&t=2693s.

13 *$19,269*: ProPublica Nonprofit Explorer, "West Oakland Environmental Indicators Project," https://projects.propublica.org/nonprofits/organizations/202384563.

14 *the city council knew this*: Bruce Hamilton, "We're Moving to Oakland, California," *Sierra*, April 6, 2016, https://www.sierraclub.org/sierra/2016-3-may-june/editor/were-moving-oakland-california.

14 *"The whole point is"*: Zennie Abraham, "Phil Tagami Day in California and Oakland," YouTube video, uploaded by Zennie62 on May 25, 2014, 2:02, https://www.youtube.com/watch?v=eJPe-JR2O70.

15 *toxic chemicals permeating the soil*: "Superfund Site: AMCO Chemical, Oakland, CA," https://cumulis.epa.gov/supercpad/cursites/csitinfo.cfm?id=0905334.

18 *the deal they cut*: Port of Oakland, "Vision 2000: Maritime Development Program," https://www.portofoakland.com/files/pdf/maritime/mari_vision.pdf.

19 *closed in the 1990s*: Bancroft Library Regional Oral History Office and Martin Meeker, "Army Base Oral Histories," 2010, https://www.loc.gov/item/2010283693/.

2. LIQUID MODERNITY

23 *"total logistics"*: John A. Grygiel, "AST&T in the Next 25 Years," *Transportation Journal* 11, no. 2 (1971): 5–13, https://www.jstor.org/stable/20712181?read-now=1&seq=1#page_scan_tab_contents.

23 *agonizing protocol meetings*: If you are a masochist and would like to read about these meetings, see Marc Levinson, *The Box: How the Shipping Container Made the World Smaller and the World Economy Bigger* (Princeton: Princeton University Press, 2016), chapter 7.

23 *"FIGURE 1. CONTAINER"*: Joint Logistics Review Board Washington D.C., "Containerization," December 1, 1969, https://apps.dtic.mil/sti/tr/pdf/AD0877965.pdf, 8.

24 *"Containerization is the technological underpinning"*: Herb Mills, interview by Chris Carlsson and Steve Stallone, *Shaping San Francisco*, April 28, 2004, https://archive.org/details/ssfHMCONTAN.

24 *contents of the fallen boxes*: U.S. Department of Commerce, "The Containerized Shipping Industry and the Phenomenon of Containers Lost at Sea," March 2014, https://nmssanctuaries.blob.core.windows.net/sanctuaries-prod/media/archive/science/conservation/pdfs/lostcontainers.pdf.

24 *More than a million boxes*: The technical measurement is the TEU, or twenty-foot equivalent. There are roughly 2.5 million of those passing through Oakland, but most boxes now are forty feet, not twenty as in the old days. So, more than a million feels like a good estimate, the real, actual number of boxes being unknown.

25 *"Living under liquid modern conditions"*: Zygmunt Bauman, *Liquid Modernity* (Cambridge, MA: Polity, 2000).

27 *roughly 85 percent of it*: UNCTAD, "Consolidation and Competition in Container Shipping," *Review of Maritime Transport 2022*, https://unctad.org/system/files/official-document/rmt2022ch6_en.pdf.

3. UNDER ALL IS THE LAND

29 *required new ideas*: Paleontologists, not least in the American Plains, had pulled monstrosities from the layers of soil/time, not just dinosaurs but creatures that looked like bizarre versions of animals still living: elephants with four tusks, long-necked rhinos, an enormous armadillo.

29 *plants that* should *rule a region*: James E. Cook, "Implications of Modern Successional Theory for Habitat Typing: A Review," *Forest Science* 42, no. 1 (1996): 67–75, https://www.uwyo.edu/vegecology/pdfs/Readings/Cook_ImplicationsofModern[1].pdf.

30 *there is Fred Clements*: Robert E. Park and Ernest W. Burgess, *Introduction to the Science of Sociology* (Chicago: University of Chicago Press, 1921), https://www.gutenberg.org/files/28496/28496-h/28496-h.htm.

30 *"In the expansion of the city"*: Ernest W. Burgess, "Residential Segregation in American Cities," *Annals of the American Academy of Political & Social Science* 140, no. 1 (1928): 105–15, https://journals.sagepub.com/doi/abs/10.1177/000271622814000115.

30 *"natural economic and cultural groupings"*: Ernest W. Burgess, "The Growth of the City: An Introduction to a Research Project," from Robert Park et al., *The City* (1925), in *The Urban Geography Reader*, ed. Nick Fyfe and Judith Kenny (London: Routledge, 2005), 19–27, https://www.taylorfrancis.com/chapters/edit/10.4324/9780203543047-4/growth-city-introduction-research-project-ernest-burgess.

31 *"This chart"*: Ibid.

31 *"Race prejudice alone"*: Burgess, "Residential Segregation in American Cities."

31 how cities worked, *generally*: Clements's ecological ideas had mostly fallen to the scientific wayside by the middle of the century. Groups of plants were not really

superorganisms. There was nothing inevitable about the progression of a region's flora. Even the basic idea that there was such thing as "formations," which became analogous to communities in sociology, had its statistical validity challenged. It's not that Clements or ecology went away, of course, but the specific features of the science underwent understandably heavy revision as new research was completed.

32 *one ethnic group or another*: The sociologists were not alone in their views on race, of course. In the mid-1920s, the National Association of Real Estate Brokers updated its code of ethics to include Article 34, an explicit prohibition on introducing different ethnic groups into new neighborhoods. Real estate brokers, the code read, "should never be instrumental in introducing into a neighborhood a character of property or occupancy, members of any race or nationality, or any individuals whose presence will clearly be detrimental to property values in that neighborhood."

32 *Burgess's key collaborator*: "Robert Ezra Park," American Sociological Association, https://www.asanet.org/robert-e-park/.

32 *"marginalized" communities*: Robert E. Park, "Human Migration and Marginal Man," *American Journal of Sociology* 33, no. 6 (1928): 881–93, https://www.jstor.org/stable /2765982.

32 *"human ecology"*: Robert E. Park, "The Concept of Position in Sociology," 1925 Presidential Address, American Sociological Association, https://www.asanet .org/wp-content/uploads/savvy/images/asa/docs/pdf/1925%20Presidential%20 Address%20(Robert%20Park).pdf.

32 *Black people were far more segregated*: Douglas S. Massey and Nancy A. Denton, *American Apartheid: Segregation and the Making of the Underclass* (Cambridge, MA: Harvard University Press, 2003), 32–33.

33 *cities did not* actually *grow in rings*: Even just the whole idea of the rings and all that— the zonal model, as it was known . . . Well, one Burgess *supporter* even had to admit at a 1939 conference: "The most widespread criticism of the Burgess zonal hypothesis arises from the fact that various cities do not actually conform to an ideal circular spatial pattern." See James A. Quinn, "The Burgess Zonal Hypothesis and Its Critics," *American Sociological Review* 5, no. 2 (1940): 210–18, https://www.jstor.org /stable/2083636.

33 *"an engine, not a camera"*: Donald MacKenzie, *An Engine, Not a Camera: How Financial Models Shape Markets* (Cambridge, MA: MIT Press, 2006).

34 *"well known to Oakland realtors"*: "Real Estate Expert to Talk," *Oakland Tribune*, February 6, 1934, 16.

34 *"an incisive analysis of hazard"*: "'Risk Rating' Styles Please," *Oakland Tribune*, October 14, 1934, 57.

34 *"the collection and tabulation"*: Ibid.

35 *months of classes*: "Realty Class to Take Test," *Oakland Tribune*, February 17, 1935, 75; also "Five Classes to Study Realty," September 1, 1935, 29; "17 Make Study of Appraisals," September 15, 1935, 28; "Students Enroll in Real Estate Classes," November 24, 1935, 33; "Appraisers Hold Meet Here," December 22, 1935, 28; "Appraisal to Be Studied," December 27, 1936, 3.

35 *"underwriting grids"*: Federal Housing Administration, *Underwriting Manual: Underwriting and Valuation Procedure Under Title II of the National Housing Act* (Washington, DC: U.S. Government Printing Office, April 1936), http://hdl.handle.net/2027 /mdp.39015018409246.

35 *a deal killer*: Jennifer Light, "Discriminating Appraisals: Cartography, Computation, and Access to Federal Mortgage Insurance in the 1930s," *Technology and Culture* 52, no. 3 (2011): 485–522, https://www.jstor.org/stable/23020643.

36 *"can scarcely be over-emphasized"*: Works Progress Administration, "1936 Real Property

Survey," Works Progress Administration Project Number 2309, Volume 1: Analyses and Summary of Data (Oakland, 1937), 11.

36 *stacks of maps*: In "Discriminating Appraisals," Jennifer Light describes the general process, and I was able to view the maps myself for Oakland at the NARA repository in College Park, Maryland.

36 *spatial technology*: Light, "Discriminating Appraisals."

36 *"There are of course some districts"*: Works Progress Administration, "1936 Real Property Survey," 13–14.

37 *"living conditions in the dwellings"*: Ibid., 51.

37 *"The housing advantage"*: Ibid., 55.

37 *"Infiltration of colored residents"*: T-RACES (Testbed for the Redlining Archives of California's Exclusionary Spaces), Oakland D-5 data, http://t-races.net/T-RACES/data /oak/ad/ad0112.pdf.

38 *Archie Williams*: Department of Commerce, Bureau of the Census, "Archie Williams U.S. Census Record," in Sixteenth Census of the United States: 1940, https://www .dropbox.com/scl/fi/3sutfd6s9ozxnt4b4ftyk/Archie-Williams-Census-record.jpg ?rlkey=vov2w3lbmdw8g2uhsy444e75b&dl=0.

39 *"I have traveled all over"*: "Williams, World Champ, Welcomed Home," *Oakland Tribune*, September 25, 1936, 39.

39 *born in Oakland in 1915*: This pocket history is drawn from Archie F. Williams, "The Joy of Flying: Olympic Gold, Air Force Colonel, and Teacher," interview by Gabrielle Morris, 1992 (Berkeley: Regents of the University of California, 1993), http:// texts.cdlib.org/view?docId=kt0v19n496&doc.view=entire_text.

4. THE DOCKS

41 *"You walk into Pier Fifteen"*: Eric Nelson recording by his nephew Brian Nelson, in the private collection of tapes created by Brian Nelson. Nelson and I listened to these tapes together, as described in the Note on Sources in this book. These half a dozen sessions deeply informed this section of the book.

42 *constant threat of serious injury*: For example, see Amy Lettman's "Disabling Injuries in Longshore Operations," *Monthly Labor Review* 112, no. 10 (1989): 37–40.

42 *down at the docks*: Herb Mills, "The Hiring Hall," FoundSF.org, http://www.foundsf .org/index.php?title=The_Hiring_Hall.

42 *crushed in 1919*: Fred Glass, *From Mission to Microchip: A History of the California Labor Movement* (Berkeley: University of California Press, 2016), 229–46.

43 The Waterfront Worker: "Wage Cuts and the Waterfront Worker," *Waterfront Worker* 1, no. 1 (December 1932), https://depts.washington.edu/labpics/zenPhoto /albums/The-ILWU-and-Longshore-Workers/The-Waterfront-Worker-complete -collection-1932-1936-new-album/1932-1933/WW%201932-1933_Page_001.jpg.

43 *a new ILA local*: Glass, *From Mission to Microchip*.

43 *a more radical direction*: U.S. Department of the Interior, National Park Service, "National Register of Historic Places Continuation Sheet, Port of San Francisco Embarcadero Historic District," January 2006, https://web.archive.org /web/20180202103512/https://sfport.com/ftp/uploadedfiles/about_us/divisions /planning_development/EmbarcaderoRegisterNominationSec8.pdf.

43 *"When one port is on strike"*: ILWU, "Remembering Bloody Thursday: July 5, 1934, on the San Francisco Waterfront," August 11, 2020, https://www.ilwu.org /remembering-bloody-thursday-july-5-1934-on-the-san-francisco-waterfront/.

44 *"Our union means a new deal"*: Bruce Nelson, *Divided We Stand: American Workers and the Struggle for Black Equality* (Princeton: Princeton University Press, 2001), 96.

44 *one would be white, and the other Black*: Both Michael Vawter and Frank Silva, from Local 10 and Local 34, respectively, cited this story to me.

45   *"Don't think of this as a riot"*: The City Museum of the City of San Francisco, "Bloody Thursday," from *San Francisco Chronicle*, July 6, 1934, http://www.sfmuseum.org /hist/thursday2.html.

45   *"The sound of thousands of feet"*: Bruce Nelson, "The Big Strike," in *Working People of California*, ed. Daniel Cornford (Berkeley: University of California Press, 1995), 225–64, https://publishing.cdlib.org/ucpressebooks/view?docId=ft9x0nb6fg&chunk .id=d0e7717&toc.id=&brand=ucpress.

45   *more than 100,000 workers struck*: "National Register of Historic Places Continuation Sheet, Port of San Francisco Embarcadero Historic District," January 2006, https:// web.archive.org/web/20180202103512/https://sfport.com/ftp/uploadedfiles/about _us/divisions/planning_development/EmbarcaderoRegisterNominationSec8.pdf.

46   *The hiring hall, jointly managed*: Mills, "The Hiring Hall."

46   *"The hiring hall was indeed the union"*: Ibid.

5. THE CALL OF CALIFORNIA

47   *"The crowd went home"*: "Arkansas Mob Lynches Negro," *Marshall News Messenger*, September 16, 1932.

47   *town of Crossett*: "Crossett Lumber Company," *Encyclopedia of Arkansas*, https:// encyclopediaofarkansas.net/entries/crossett-lumber-company-3588/, updated January 5, 2024.

47   *the company-owned store*: Don C. Bragg, "A Brief History of Forests and Tree Planting in Arkansas," archived at https://web.archive.org/web/20170216051513/https:// www.srs.fs.fed.us/pubs/ja/2012/ja_2012_bragg_003.pdf.

48   *the faces might have been smiling*: For the most evocative treatment of the cruelty and strangeness of the white perspective of lynching, the artist Ken Gonzalez-Day's *Erased Lynching* highlights the perpetrators while allowing the victims of this racial violence a type of dignity.

48   *headed the household*: Department of Commerce, Bureau of the Census, "Clemmie in Arkansas," Sixteenth Census of the United States: 1940, https://www.dropbox .com/scl/fi/008r72mhyo70kpc84q461/Clemmie-in-Arkansas.jpg?rlkey=spsx266 taxmlrxetovk63no8g&dl=0.

48   *Of the men on the block*: David M. Moyers, "Trouble in a Company Town: The Crossett Strike of 1940," *Arkansas Historical Quarterly* 48, no. 1 (1989): 34–56, https://www.jstor .org/stable/40027805.

48   *their daddy working in the sawmill*: U.S. Census, April 26, 1940, Arkansas, Ashley, Crossett, Egypt Township; ED 2-10, sh 28B; lines 41–80.

48   *An exodus*: Donald Holley, "Leaving the Land of Opportunity: Arkansas and the Great Migration," *Arkansas Historical Quarterly* 64, no. 3 (2005): 245–61, https://web .archive.org/web/20120916082608id_/http://faculty.nwacc.edu/dvinzant/documents /DonaldHolley.pdf.

48   *"I'm a tell you about San Francisco"*: *Take This Hammer* (the Director's Cut), directed by Richard O. Moore (San Francisco: KQED Film Unit, 1963), Bay Area Television Archive, https://diva.sfsu.edu/collections/sfbatv/bundles/216518.

49   *"in trailers, tents, tin houses"*: United States Congress House Committee on Naval Affairs, "Investigation of Congested Areas: San Francisco, Calif., Area," April 12–17, 1943 (Washington, DC: U.S. Government Printing Office, 1943), 867.

49   *The music was good, though*: Shirley Ann Wilson Moore, *To Place Our Deeds: The African American Community in Richmond, California, 1910–1963* (Berkeley: University of California Press, 2012), 131. The musician Jimmy McCracklin immortalized one spot in his lyrics to "Club Savoy": "Now, Club Savoy's where the fine chicks make it / From eight to eighty, triple cross-eyed blind and crazy / If you ever go there and you want to shake your fanny / If things don't go right, just call Granny!"

49 *In one photo*: This photo can be found in *Guide to Federal Records in the National Archives of the United* States, compiled by Robert B. Matchette et al. (Washington, DC: National Archives and Records Administration, 1995). See Records of the United States Maritime Commission, 178.14. Records of the West Regional Office, consisting of photographs and negatives of the Richmond Housing Project, 1942–43 (in San Francisco).

50 *only a quarter of the time*: City Manager of Richmond, California, "An Avalanche Hits Richmond," July 1944. See Community Archive, *How the Bay Was Built 9*, https://howthebaywasbuilt.com/richmond/.

50 *In a photo by Dorothea Lange*: Dorothea Lange, *Richmond—School Children—Every Hand Up Signifies a Child Not Born in California*, Richmond, CA, circa 1942, The Dorothea Lange Collection at the Oakland Museum of California, https://dorothealange.museumca.org/image/richmond-school-children-every-hand-up-signifies-a-child-not-born-in-california/A67.137.42115.1/.

50 *made the best of it*: "Rosie the Riveter/WWII Home Front National Historical Park, California," Museum Collections, U.S. National Park Service, https://museum.nps.gov/ParkPagedet.aspx?rID=RORI%20%20%20%20%20%20%201%26db%3Dobjects%26dir%3DPARKS.

50 *In another Lange photo*: Dorothea Lange, *Untitled (Shipyard Worker)*, Richmond, CA, circa 1943, The Dorothea Lange Collection at the Oakland Museum of California, https://dorothealange.museumca.org/image/untitled-shipyard-worker/A67.137.42058.1/.

50 *"Hitler was the one"*: Fanny Christina Hill, "'Hitler Was the One That Got Us Out of the White Folks' Kitchen,'" *The Columbia Documentary History of American Women Since 1941*, ed. Harriet Sigerman (New York: Columbia University Press, 2003), 35–41.

50 *part of the Messmen*: Naval History and Heritage Command, "The Negro in the Navy," https://www.history.navy.mil/research/library/online-reading-room/title-list-alphabetically/n/negro-navy-1947-adminhist84.html.

51 *"the enlistment of Negroes"*: Ibid.

51 *his mother, Lela Mae*: "Mrs. Lela Hunter Funeral at Anson," *Abilene Reporter-News*, September 2, 1958, 25.

51 *"undermine the value"*: City Manager of Richmond, "An Avalanche Hits Richmond," 12.

51 *scheduled for destruction*: Peter P. F. Radkowskii III, "Managing the Invisible Hand of the California Housing Market, 1942–1967," *California Legal History Journal* 1 (2006): 32, https://www.law.berkeley.edu/wp-content/uploads/2015/04/radkowski-paper.pdf.

6. SPOILED BY THE VARIOUS MUTATIONS OF THE AIR

53 *"Why should I fight"*: "Jessica Mitford Part 4 of 8," *FBI Records: The Vault*, Decca Treuhaft FBI file, part 4, p. 13, https://vault.fbi.gov/jessica-mitford/jessica-mitford-part-4-of-8/view.

53 *"For the first time"*: Mabel L. Walker, *Urban Blight and Slums* (Cambridge, MA: Harvard University Press, 1938), 193.

53 *"spoiled by the various mutations of the Air"*: J. W. Gent [John Worlidge], *Systema agriculturæ; the mystery of husbandry discovered . . . The third edition . . . with one whole section added, and many large and useful additions* (Tho. Dring, 1681), 159.

53 *"anything which withers"*: "Blight" in *Oxford English Dictionary*.

54 *"Negroes are being discriminated against"*: "U.S. Housing Project Hit by 40 Property Owners," *Oakland Tribune*, April 14, 1939, 19.

54 *didn't quite stick*: For the broader political context of this battle, see chapter 8, "Boom-

town Blues," of Marilynn S. Johnson, *The Second Gold Rush: Oakland and the East Bay in World War II* (Berkeley: University of California Press, 1996).

54 *"a serious and growing menace"*: 1945 Community Redevelopment Act.

54 *Supreme Court*: This was the Warren Court. Just a few months earlier they'd decided to outlaw school segregation. Coincidentally (or is it?), Warren had been the district attorney in Alameda County, and a crucial part of the Athens Club–belonging Republican power structure in Oakland.

55 *"Miserable and disreputable housing conditions"*: Berman v. Parker, 348 U.S. 26 (1954), Justia U.S. Supreme Court Center, https://supreme.justia.com/cases/federal/us/348 /26/.

55 *"less worthy"*: Wendell E. Pritchett, "The 'Public Menace' of Blight: Urban Renewal and the Private Uses of Eminent Domain," *Yale Law & Policy* 21 (2003): 2.

56 *"penalty score"*: City Planning Commission, Oakland, California, "Oakland Area Residential Analysis," July 1956, https://www.dropbox.com/scl/fi/46l5abn7hrrmgz3bc33 qm/1956-Oakland-Residential-Area-Analysis.pdf?rlkey=b6gw55oud9aado50i1n1d v6nd&dl=0.

56 *a myriad of weighted calculations*: See Appendix A of Residential Area Analysis.

56 *"A total of 1,140 acres"*: *Oakland Tribune*, July 22, 1956, 1.

56 *the blighted area*: See Alexis C. Madrigal, "Oakland Residential Area Analysis," available online at https://docs.google.com/spreadsheets/d/1LUe3kIlgqDS1Cc9DLnM6 A4jTCueVtO34vsbRbMmp_mA/edit?usp=sharing.

57 *new post office distribution center*: *Oakland Tribune*, October 24, 1958.

57 *"rise in the values"*: Homer Hoyt, *One Hundred Years of Land Values in Chicago* (Chicago: University of Chicago Press, 1933), 317, https://archive.org/details/onehundred yearso00hoytrich/page/316/mode/2up?q=outer+areas.

58 *sixteen ships at a time*: National Park Service, "San Francisco Port of Embarkation: Gateway to the Pacific," https://www.nps.gov/articles/sanfrangateway.htm, updated November 9, 2015.

58 *reshaping the bay*: All told, one sixth of the city's area had once been tideland. See Oakland City Planning Commission, "Shoreline Development Plan," 1952, https:// howthebaywasbuilt.com/oakland/#1952shoreline.

58 *the water surrounding Oakland*: Board of Consulting Engineers et al., "East Bay Cities Sewage Disposal Survey," June 30, 1941, 57, https://rfs-env.berkeley.edu/sites /default/files/publications/1941.06.30eastbaysewagedisposalsurvey.pdf.

58 *zoned for industrial use*: Oakland (Calif.) City Planning Commission, *Zoning Laws, Oakland, California. Ordinance 474–475 C.M.S., Adopted February 5, 1935. Text of Ordinances and Maps, rev. to March 31, 1946* (Oakland: Tribune Press, 1946), https://babel .hathitrust.org/cgi/pt?id=uc1.b2657274&seq=35.

58–59 *"The industrialization of China"*: Ibid., 27.

59 *"These lands"*: Ibid., 28.

59 *NSC 48 and 48/2*: U.S. Department of State Office of the Historian, "Memorandum by the Executive Secretary of the National Security Council (Souers) to the National Security Council," in *Foreign Relations of the United States, 1949, The Far East and Australasia* VII, Part 2, ed. J. G. Reid and J. P. Glennon (Washington, DC: United States Government Printing Office, 1976), https://history.state.gov/historicaldocuments /frus1949v07p2/d387.

60 *15 percent of the population*: Charles K. Armstrong, "The Destruction and Reconstruction of North Korea, 1950–1960," *Asia-Pacific Journal* 7, no. 0 (2009): https://apjjf .org/charles-k-armstrong/3460/article.

60 *laying waste to the country*: Bruce Cumings, *The Korean War: A History* (New York: Modern Library, 2010), 186, and Taewoo Kim, "Limited War, Unlimited Targets:

U.S. Air Force Bombing of North Korea During the Korean War, 1950–1953," *Critical Asian Studies* 44, no. 3 (2012): 467–92, https://doi.org/10.1080/14672715.2012.711980.

60 *"We killed civilians"*: Karig, 1952, quoted in Cumings, *The Korean War.*

60 *their parents dead*: Taewoo Kim, "Overturned Time and Space: Drastic Changes in the Daily Lives of North Koreans During the Korean War," *Asian Journal of Peacebuilding* 2, no. 2 (2014): 241–62, https://web.archive.org/web/20150701203519/http://s-space.snu.ac.kr/bitstream/10371/93639/1/07_Taewoo%20Kim.pdf.

60 *"I have seen"*: Quoted in War Is Boring, "How the Korean War's Brutality Turned the Stomachs of America's Most Hardened Soldiers," *National Interest*, April 13, 2019, https://nationalinterest.org/blog/buzz/how-korean-war%E2%80%99s-brutality-turned-stomachs-america%E2%80%99s-most-hardened-soldiers-52237.

61 *"It was on this hauntingly beautiful island"*: Cumings, *The Korean War*, 164.

61 *Military spending quadrupled*: Ibid.

61 *15 percent of American GDP*: Maurice R. Greenberg, "Trends in U.S. Military Spending," *Council on Foreign Relations*, July 15, 2014, https://files.cfr.org/sites/default/files/pdf/2012/08/Trends%20in%20US%20Military%20Spending%202014_final.pdf.

61 *"perfect laboratory"*: Margaret O'Mara, *The Code: Silicon Valley and the Remaking of America* (New York: Penguin, 2019), 30.

62 *radars, missiles, and bombs*: Ibid., chapter 2. See also Malcolm Harris, *Palo Alto* (New York: Little, Brown, 2023).

7. ONE WAS DOWN, ONE WAS UP

63 *That's 0.029 percent*: Jane Kim, "Black Reparations for Twentieth Century Federal Housing Discrimination: The Construction of White Wealth and the Effects of Denied Black Home Ownership," *Boston University Public Interest Law Journal* 29, no. 135 (2019): 23, https://papers.ssrn.com/sol3/papers.cfm?abstract_id=3615775.

63 *a determinative role*: For a multitude of Bay Area examples, see Richard Rothstein, *The Color of Law: A Forgotten History of How Our Government Segregated America* (New York: Liveright, 2017), chapter 1.

63 *"The long-established ideals"*: Richmond City Council, "The Master Plan of Richmond California," 1950, https://howthebaywasbuilt.com/wp-content/uploads/2019/01/1950-richmond-housing-and-redevelopment_complete.pdf, 20–21.

64 *"At this period"*: U.S. Federal Housing Administration and Homer Hoyt, *The Structure and Growth of Residential Neighborhoods in American Cities* (Washington, DC: U.S. Government Printing Office, 1939), 121.

64 *all-white Rollingwood development in 1952*: See Rothstein, *The Color of Law*, chapter 9.

64 *they found a white crowd*: Jessica Mitford, *A Fine Old Conflict* (New York: Random House, 1978), 128–29. We have this remarkable account thanks to Jessica Mitford's memoir. Mitford, known then as Decca Treuhaft, was the head of the Civil Rights Congress in the East Bay. The CRC was a Communist front group and a dedicated racial justice institution, battling segregation, police brutality, and other forms of discrimination. Mitford was a British aristocrat, made famous because her family primarily went fascist while she went Communist, running off to the Spanish Civil War. She landed in Oakland and met the locally legendary lawyer Bob Treuhaft. They threw themselves into work for the Party—Decca answering both to the Party and to the largely Black board of the CRC.

64 *"organizing patrols"*: Part 4 of the Mitford/Treuhaft FBI file: https://vault.fbi.gov/jessica-mitford/jessica-mitford-part-4-of-8/view.

64 *in those days*: Mitford, *A Fine Old Conflict*.

65 *They had white neighbors, too*: 1950 Federal Census.

66 *"completely unsuited for residential use"*: Oakland City Planning Commission, "Shoreline Development Plan," 1952, 18, 54.

67 *a future of racial harmony*: For a deep investigation of West Oakland's long history, see "Putting the 'There' There: Historical Archaeologies of West Oakland," *Cypress Replacement Project Interpretive Report No. 2*, ed. M. Praetzellis and A. Praetzellis (June 2004), https://asc.sonoma.edu/sites/asc/files/praetzellis_2004_putting_there_there_west_oakland.pdf.

67 *just one block down*: See *Evolutionary Blues . . . West Oakland's Music Legacy*, directed by Cheryl Fabio (Oakland: KTOP, 2017).

67 *"Slim Jenkins"*: Harold Jenkins, "Slim Jenkins business card," African American Museum and Library at Oakland, https://calisphere.org/item/0886a1b4532a118b318882b36f23664f/.

67 DAISY BRIDGE CLUB: E. F. Joseph, "Women Seated at White-Cloth Covered Tables in Slim Jenkins Bar and Restaurant . . . ," African American Museum and Library at Oakland, https://calisphere.org/item/7fb7f7f500d1018cf78519a9dc5863dc/.

67 *Sportland Recreation Hall*: "Poster for Mary Wells Performing at Esther's Orbit Room," https://projects.journalism.berkeley.edu/7thstreet-archive/category/sources/other/.

67 *"know the score"*: Berkeley journalism students assembled a list of the run-ins: https://projects.journalism.berkeley.edu/7thstreet-archive/category/sources/.

67 *Police maintained there was a dice game*: Raincoat Jones, *Oakland Tribune*. Also http://kintecenter.org/PAGES/seventhst.html.

68 *the segregated economy*: Gretchen Lemke-Santangelo, *Abiding Courage: African American Migrant Women and the East Bay Community* (Chapel Hill: University of North Carolina Press, 1996), chapter 4.

68 *"I would get jobs for people"*: Esther Mabry, "Oral History Interview with Esther Mabry," interview by African American Museum and Library at Oakland, 2002, https://californiarevealed.org/do/97295ae0-188e-425d-96ca-264375d3b633.

69 *the bartender Ted Jernigan*: The legendary Betty Reid Soskin certainly remembered him; see Betty Reid Soskin, "And Still Later . . . ," May 13, 2007, http://cbreaux.blogspot.com/2007_05_13_archive.html.

8. KILLING YOU WITH THAT PENCIL AND PAPER

70 *watching the destruction*: This was all detailed in the *Oakland Tribune*, August 16, 1960, 13.

70 *"The Battle for Oakland"*: For an example of the AP treatment, see *Manitowoc Herald-Times*, August 16, 1960, 28.

71 *automation to the USPS*: *Oakland Tribune*, August 30, 1959, 16.

71 *"mostly substandard housing"*: "Mail Center Brings Home Problems," *Oakland Tribune*, August 27, 1959, 1.

71 *She'd earned a doctorate in race relations*: See David McEntire, *Where Shall We Live? Report of the Commission on Race and Housing* (Berkeley: University of California Press, 2023).

72 *obligated by law to create*: *Oakland Tribune*, September 3, 1959.

72 *paper pushers of one kind or another*: "The Coming Victory Over Paper," Edmund L. Van Deusen, cited in https://archive.org/details/LetErmaDoIt.

72 *"Sales slips, purchase orders, payrolls"*: David Oakes Woodbury, *Let Erma Do It. The Full Story of Automation* (New York: Harcourt, Brace, 1956).

72 *along with a billion parcels*: Arthur E. Summerfield, *U.S. Mail: The Story of the United*

States Postal Service (New York: Holt, Rinehart, and Winston, 1960), 10–11, https://archive.org/details/usmail0000unse/page/10/mode/2up?q=half+the+citizens.

72 *"half the citizens"*: United States Congress Senate Committee on Post Office and Civil Service, *Explanation of the Postal Reorganization Act and Selected Background Material* (Washington, DC: U.S. Government Printing Office, 1975), 286.

73 *"revolutionary post offices"*: Summerfield, *U.S. Mail*, 179–81.

73 *a dismal failure*: Nancy Pope, "Project Turnkey—Providence, RI," Blog of the Smithsonian National Postal Museum, October 22, 2018, https://postalmuseum.si.edu/project-turnkey-%E2%80%93-providence-ri.

74 *waves of longshoremen*: Charles Snipes, "Oakland Army Base Oral History Project," interview by Jess Rigelhaupt, 2008 (Oral History Center, Bancroft Library, University of California, Berkeley, 2009), https://digicoll.lib.berkeley.edu/record/218517?ln=en.

74 *all managed on paper*: U.S. Office of Naval Research, *Naval Logistics Research Quarterly*, vols. 4 and 5 (Washington, DC: Office of Naval Research, 1957), 228.

74 *"A superficial view"*: Ibid.

74 *their new machine*: Kenneth Flamm, *Creating the Computer: Government, Industry and High Technology* (Washington, DC: Brookings Institution, 2010), 69.

75 *increasingly complex logistics simulations*: Murray A. Geisler, "A First Experiment in Logistics System Simulation," *Naval Research Logistics Quarterly* 7 (1960): 21–44, https://onlinelibrary.wiley.com/doi/abs/10.1002/nav.3800070105.

75 *"Cargo handling has remained"*: University of California Los Angeles, Department of Engineering, "An Engineering Analysis of Cargo Handling" (October 1953), 1, https://apps.dtic.mil/sti/tr/pdf/AD0020834.pdf.

75 *rapidly scaling up*: This story was told well by the ILWU in Louis Goldblatt, *Men and Machines* (San Francisco: International Longshoremen's & Warehousemen's Union, 1963).

75 *preserved their bodies*: Lincoln Fairley, "The ILWU-PMA Mechanization and Modernization Agreement," *Labor Law Journal* (1961): 664–80, https://oac.cdlib.org/ark:/28722/bk0003z5f46/?brand=oac.

76 *turn a ship around*: United States Interstate Commerce Commission, *Reports and Decisions of the Interstate Commerce Commission of the United States* 195 (Washington, DC: U.S. Government Printing Office, 1934), 215–34.

77 *"a complete scheme"*: United States Army Quartermaster Corps, *Water Transportation* (Washington, DC: U.S. Government Printing Office, 1941), 71.

77 *Alaska Steamship*: Jennifer Ott, "First Modern Container Ship to Sail from Puget Sound Departs the Port of Seattle for Hawaii on January 24, 1964," HistoryLink.org, September 9, 2014, https://www.historylink.org/File/10923.

77 *Implementing these systems*: See Levinson, *The Box*, 51, among other sections.

9. RIGHT-OF-WAY MEN

78 *"a second Manhattan"*: Burton H. Wolfe, "BART: Bechtel's Baby," FoundSF.org, originally published in the *San Francisco Bay Guardian*, February 14, 1973, https://www.foundsf.org/index.php?title=BART%3A_Bechtel%27s_Baby.

79 *the Bechtels*: United States Congress Office of Technology Assessment, *Assessment on Community Planning for Mass Transit: Volume 8—San Francisco Case Study* (February 1976), https://www.dropbox.com/scl/fi/6ts7rhijj02ak8b7v88ae/Assessment-of-Community-Planning-for-Mass-Transit-Volume-8-San-Francisco-Case-Study.pdf?rlkey=9tblaozyy5cckwpljhxmz6d6g&dl=0.

79 *"In 100 Years We've Moved 10 Feet"*: Much of the flavor of the BART planning process is drawn from Wally Mersereau's collection of BART planning documents and brochures, held in the San Francisco History Center: *Wallace D. Mersereau San Fran-*

cisco Bay Area Rapid Transit (BART) District Real Estate Records, https://oac.cdlib.org
/findaid/ark:/13030/c8zs32p2/entire_text/.

79 *"the largest single public works project"*: San Francisco Bay Area Rapid Transit
System, "A History of BART: The Concept Is Born," https://www.bart.gov/about
/history.

79 *"The Future Is Now"*: Stokes's series ran in the *Oakland Tribune*, May 9–18, 1954.

79 *"cohesive metropolitan area"*: Oakland Tribune, May 16, 8.

79 *"de-centralization"*: From the May 9 introductory article in the series, 7.

80 *But they did*: Early Days in BART Real Estate (1963–1973), Wallace Mersereau, 2016;
Box 1, https://oac.cdlib.org/findaid/ark:/13030/c8zs32p2/entire_text/.

80 *extending its tentacles in California*: Elisa Barbour and Paul G. Lewis, "California
Comes of Age: Governing Institutions, Planning, and Public Investment," Public
Policy Institute of California, June 2, 2005, 7–9, https://www.dropbox.com/scl/fi
/2agv2y765dbflzt1hh0n7/California_Comes_of_Age_Governing_Instit.pdf?rlkey
=a823mtymdkd6vtrpr0tq9y35c&dl=0.

80 *"Engineer's Answer"*: "California Public Works," *California Highways and Public Works*
37, issues 1–2 (January–February 1958): 26, https://cdm16436.contentdm.oclc.org
/digital/collection/p16436coll4/id/13903/rec/311.

80 *"13 Million People Want My Property"*: California Department of Highways, "13
Million People Want My Property" (1956 pamphlet), https://www.dropbox
.com/scl/fi/yo038a0s8t9jb75zlx51f/13-million-people-want-my-home.pdf?rlkey
=sgn0wdu7vvxqhrcedy1qby33i&dl=0.

81 *97.4 percent white in 1960*: Census Data, http://www.bayareacensus.ca.gov/cities
/DalyCity50.htm. Mersereau recalled Daniels fondly. And Daniels was not without
his peccadilloes acquired over the decades. "He never tapped his cigarette on an ash-
tray to flick off the ash, but handled the cigarette gently while keeping an eye on it
as the ash slowly lengthened," a subordinate remembered. "When the ash neared its
potentially greatest length it would fall. The long ash falling on the front of his dark
suit was automatically followed by a brushing motion with his nonsmoking hand to
try to remove the ash."

81 *"We liked to remind ourselves"*: The office seems to have been rich in mid-century
camaraderie, according to the mini-memoir of Wally Mersereau, who was one of
Daniels's guys. Friday mornings found the men enjoying "Bagel Day," courtesy of
Irv Pilch, who Mersereau remembered "had a source of bagels." Some office joker
"made a small pennant on a stick that was fastened to the radiator and was dubbed
'the bagel flag.'" When the bagels came in, the flag went up, and word spread round
the office, men in suits warming their bagels on a radiator, talking shop.

81 *"deal with individual property owners"*: Wally Mersereau collection, like much of this
narrative.

83 *began to sell in February 1965*: This information is all extracted from BART's
real estate records, which I acquired through extensive communication with
BART. The full set of files I received is available online: https://www.dropbox
.com/scl/fo/hhkhpxf9st3d9qchkz684/AOTMFBsTgxymggUEQ3xD7OA?rlkey
=t4ckw9pp5nqig9u2z6395d81u&dl=0.

83 *"Director Anderson, continuing"*: These documents are held in the BART executive
offices, and can be accessed though not through an official archival system.

84 *saw something else*: For more on this fight, see Joseph A. Rodriguez, "Rapid Transit and Community Power: West Oakland Residents Confront BART," *Antipode* 31,
no. 2 (1999): 212–28, and Darwin BondGraham, "JoBART: Rapid Transit, Race, and
the Construction of a Global San Francisco," https://www.scribd.com/document
/151383987/JoBART-Rapid-Transit-Race-and-the-Construction-of-a-Global-San
-Francisco.

84 *a letter from Sneed*: Some of Sneed's materials are held in the San Francisco State Labor Archives and Research Center in the Data Center Records collection, Series 8, Subseries 8.1.

84 *"The President has asked me"*: Present in the JOBART archives at San Francisco State.

84 *"[Sneed] held the council near-spellbound"*: *Oakland Tribune*, February 16, 1966, 21.

85 *printed the speech in full*: "Mrs. Sneed Goes to City Hall," *Flatlands*, March 12, 1966, 7.

85 *"The landlord stopped collecting rent"*: "The Poor Speak Out," *Flatlands*, March 12, 1966, 7.

86 *"essentially personal sacrifices"*: Anthony Downs, "Uncompensated Nonconstruction Costs Which Urban Highways and Urban Renewal Impose upon Residential Households," in *The Analysis of Public Output*, ed. Julius Margolis (Washington, DC: National Bureau of Economic Research, 1970), https://www.nber.org/system/files/chapters/c3351/c3351.pdf.

86 *population fell by 14,000*: Robert O. Self, *American Babylon: Race and the Struggle for Postwar Oakland* (Princeton: Princeton University Press, 2005), 155.

87 *"Ordinary people"*: Guy B. Adams and Danny L. Balfour, *Unmasking Administrative Evil*, 3rd ed. (New York: Routledge, 2014), 4.

87 *"condition[ed] Americans to accept racism"*: Robert C. Weaver, *The Negro Ghetto* (New York: Harcourt, Brace, 1948).

87 *"Who are we kidding"*: "The 'problem' of housing minority groups," remarks by Robert B. Pitts, racial relations officer, Federal Housing Administration, before Conference on Housing Minority Groups, Benjamin Franklin Hotel, Seattle, Washington, Box 2:2, https://oac.cdlib.org/findaid/ark:/13030/c8q52rjs/.

87 *"Quantification is a way"*: Theodore M. Porter, *Trust in Numbers: The Pursuit of Objectivity in Science and Public Life* (Princeton: Princeton University Press, 2020), 8.

88 *"The state was subsidizing"*: David M. P. Freund, *Colored Property: State Policy and White Racial Politics in Suburban America* (Chicago: University of Chicago Press, 2007), 36–37.

88 *Most of them signed away their properties*: From the BART archival deed files and maps.

88 *downtown Oakland*: BART archival deed map, available online at the Pacific Circuit website.

88 *"Through eminent domain"*: Oscar Carl Wright, "Oral History Interview with Oscar Carl Wright," interview by Barbara Cannon, 2003 (African American Museum and Library at Oakland), https://californiarevealed.org/do/6fb45820-9ee6-4b60-97a1-896efd50a650.

88 *"She loved music and traveling"*: "Beatrice M. Sneed, Obituary," *East Bay Times*, July 31, 2001, https://www.eastbaytimes.com/obituaries/beatrice-m-sneed/.

10. THE NEGRO FAMILY

89 *750 revolts*: Peter B. Levy, *The Great Uprising: Race Riots in Urban America in the 1960s* (Cambridge: Cambridge University Press, 2018).

90 *"turning point"*: Aliyah Dunn-Salahuddin, "A Forgotten Community, a Forgotten History: San Francisco's 1966 Urban Uprising," in *The Strange Careers of the Jim Crow North: Segregation and Struggle Outside of the South*, ed. B. Purnell, J. Theoharis, and K. Woodard (New York: New York University Press, 2019), 211–34.

90 *"rats and roaches"*: As noted in the Note on Sources, *The Spokesman* was a hyperlocal paper supported by War on Poverty funds. Many details of Hunters Point in the late '60s appear only in this paper, and I used it liberally throughout this section. It is now available on the Internet Archive, and an index created by my heroic mother is available at the Pacific Circuit website: https://archive.org/search?query=creator%3A%22Simms%2C+Robert%22.

90 *"It is a very shocking experience"*: Perhaps it is not surprising that in later decades, Double Rock—Ms. Margaret's exact corner, actually, Fitzgerald and Griffith—became the murder capital of San Francisco.

91 *"tangle of pathology"*: U.S. Department of Labor and Daniel Patrick Moynihan, *The Negro Family: The Case for National Action* (March 1965), https://www.dol.gov/general /aboutdol/history/webid-moynihan.

91 *"individual personality problems"*: Arthur E. Hippler, *Hunter's Point: A Black Ghetto in America* (New York: Basic Books, 1974), 6.

92 *"personal pathology"*: Ibid.

92 *"the Big 5"*: The Spokesman 1, no. 15.

93 *"If you are from Hunters Point"*: Cited in Dunn-Salahuddin, "A Forgotten Community, a Forgotten History."

93 *from Dianne Feinstein to Willie Brown*: There's a great set of clippings at the San Francisco Public Library about Eloise Westbrook's many civic engagements.

93 *"Is there any way"*: The Spokesman 3, nos. 5–6.

93 *"I've had a baby every year"*: The Spokesman 1, no. 16.

11. THE POTENTIALITIES ARE BEYOND IMAGINING

97 *"Globalization"*: Levinson, *The Box*, 18.

98 *Fairchild Semiconductor*: Jeffrey Henderson, "The Political Economy of Technological Transformation in Hong Kong," in *Pacific Rim Cities in the World Economy: Comparative Urban and Community Research*, Vol. 2, ed. Michael Peter Smith (New Brunswick, NJ: Transaction Publishers, 1989), 109.

98 *Everybody was cheaper in Hong Kong*: Christophe Lécuyer, *Making Silicon Valley: Innovation and Growth of High Tech, 1930–1970* (Cambridge, MA: MIT Press, 2007), 205.

99 *"information, energy, and materials"*: Stanford Research Institute, "The Anatomy of Production Automata," in *Symposium on Automatic Production of Electronic Equipment, Proceedings, April 19 and 20, 1954, Fairmont Hotel, San Francisco, California*, 13.

100 *"It may be true"*: National Industrial Conference Board, *Proceedings Symposium on Electronics and Automatic Production, Proceedings*, Vol. 1 (Stanford Research Institute, 1956), Woolridge address, 1.

100 *"My soul has been sucked dry"*: Gene Dennis read this poem at a Waterfront Writers and Artists event, captured on tape by Brian Nelson. See Note on Sources for more detail.

101 *The story of "a trucker"*: For example, Anthony J. Mayo and Nitin Nohria, "The Truck Driver Who Reinvented Shipping," excerpt from *In Their Time: The Greatest Business Leaders of the Twentieth Century*, Harvard Business School, October 3, 2005, https:// hbswk.hbs.edu/item/the-truck-driver-who-reinvented-shipping.

101 *He published the first paper*: Foster L. Weldon, "Cargo Containerization in the West Coast–Hawaiian Trade," *Operations Research* 6, no. 5 (1958): 649–790, https://pubsonline .informs.org/doi/abs/10.1287/opre.6.5.649.

101 *"the Big 5"*: Charlotte Curtis, "Where Aristocrats Do the Hula and Vote Republican," *New York Times*, November 7, 1966, 90.

101 *The field was flush*: M. Fortun and S. S. Schweber, "Scientists and the Legacy of World War II: The Case of Operations Research (OR)," *Social Studies of Science* 23, no. 4 (1993): 595–642, https://doi.org/10.1177/030631293023004001.

102 *labor savings on the waterfront*: Levinson, *The Box*, stuff on the 1970s.

102 *a serious check on the Big 5*: See Jung Moon-Kie, *Reworking Race: The Making of Hawaii's Interracial Labor Movement* (New York: Columbia University Press, 2006).

102 *Weldon's findings*: Oakland Tribune, September 8, 1968, 22.

102 *container-handling equipment*: Foster L. Weldon, "Cargo Containerization in the West Coast–Hawaiian Trade," *Operations Research* 6, no. 5 (1958): 649–70, https://doi.org/10 .1287/opre.6.5.649. "History: Matson's Contributions to Hawaii," Matson.com, https:// www.matson.com/corporate/about_us/history.html.

103 *within a couple of decades*: "America on the Move" exhibit, National Museum of

American History, Washington, D.C., November 22, 2023—Permanent, https://americanhistory.si.edu/america-on-the-move/transforming-waterfront.

103 *"Containers had grown so much faster"*: Ben E. Nutter, "The Port of Oakland: Modernization and Expansion of Shipping, Airport, and Real Estate Operations," interview by Ann Lage, 1991 (Berkeley: Regional Oral History Office, University of California, 1994), https://digitalassets.lib.berkeley.edu/roho/ucb/text/nutter_ben.pdf.

103 *stacked like Lego bricks*: These stories are told in Levinson's *The Box*.

104 *poet-novelist named George Benet*: All Benet stories are drawn from the Brian Nelson tape archive.

105 *"and talk about a problem"*: Nutter, "The Port of Oakland: Modernization and Expansion of Shipping, Airport, and Real Estate Operations."

105 *Then the men in the novel*: George Benet, *A Short Dance in the Sun* (Los Angeles: Lapis Press, 1988).

107 *a polarizing agreement*: See, for example, this award-winning undergraduate thesis: James Bradley, "'The Union Ruptured': Mechanization, Modernization, and the International Longshore and Warehouse Union" (BA thesis, University of California Berkeley, 2016), https://escholarship.org/content/qt4h6877f9/qt4h6877f9.pdf.

109 *This beautiful place*: These descriptions are drawn from a repository of military footage of Cam Ranh Bay facilities at CriticalPast.com. See "Aerial views of the port facilities at Cam Ranh Bay, during the Vietnam War," criticalpast.com, https://www.criticalpast.com/video/65675067521_naval-bases_naval-training-center_four-large-piers_aerial-view-of-buildings.

109 *a way station*: Marcia A. Eymann and Charles Wollenberg, eds., *What's Going On? California and the Vietnam Era* (Berkeley: University of California Press, 2004), 1–4.

109 *"We've been handling cargo"*: Nutter, "The Port of Oakland: Modernization and Expansion of Shipping, Airport, and Real Estate Operations."

109 *The Save the Bay movement*: "Our Mission," savethebay.org, https://savesfbay.org/about-us/.

109 *filled in by 2020*: *Oakland Tribune*, April 23, 1961, 69.

110 *"of such unforgettable beauty"*: Rice Odell, *The Saving of San Francisco Bay: A Report on Citizen Action and Regional Planning* (San Francisco: Conservation Foundation, 1972), xi.

110 *"We knew that if we waited"*: Nutter, "The Port of Oakland: Modernization and Expansion of Shipping, Airport, and Real Estate Operations."

110 *named after Nutter*: *Oakland Tribune*, September 8, 1968, 19.

110 *"In the next 20 years"*: *Oakland Tribune*, January 23, 1966, 173.

110 *"There were a few docks"*: Levinson told me this during an interview.

111 *"the largest military port complex"*: Oakland Seaport, "Seaport Logistics Complex History and Timeline," www.oaklandseaport.com/development-programs/seaport-logistics-complex/seaport-logistics-complex-history-and-timeline/.

111 *the 1960s' fastest-growing economy*: Levinson, *The Box*, 250.

111 *already strong connection*: Ibid., 292.

111 *half of Sea-Land's business*: Oakland Army Base Oral History, 13.

111 *insensitivity to community concerns*: For a broader history of the containerization of Oakland and specifically the role of federal funding for the Seventh Street Terminal, see Mark Rosenstein's master's thesis, "The Rise of Maritime Containerization in the Port of Oakland, 1950 to 1970" (New York University, 2000), www.apparentwind.com/mbr/maritime-writings/thesis.pdf.

112 *"a different era"*: Nutter, "The Port of Oakland: Modernization and Expansion of Shipping, Airport, and Real Estate Operations," see introduction by Walter Abernathy.

12. BLACK POWER AND THE THIRD WORLD MOVEMENT

114 *"Mau-Mauing the Flak Catchers"*: Wolfe used pseudonyms sometimes, maybe all the time, though he is not clear about this. I couldn't find any documentary evidence for the stories he told in the paper, especially about the violence and corruption of local people.

114 *philosophically disjointed*: Martha J. Bailey and Sheldon Danziger, eds., *Legacies of the War on Poverty* (New York: Russell Sage Foundation, 2013).

114 *"magnificent idea"*: Ronald V. Dellums, *Lying Down with the Lions: A Public Life from the Streets of Oakland to the Halls of Power* (Boston: Beacon Press, 2000), 35.

115 *"I thought coming to California"*: KRON-TV, "Assignment Four: Segregation Western," 1963. Archived by the Bay Area Television Archive at https://diva.sfsu.edu/collections/sfbatv/bundles/223879.

116 *this kind of solidarity*: See, for example, Luis Eslava, Michael Fakhri, and Vasuki Nesiah, "The Spirit of Bandung," in *Bandung, Global History, and International Law*, ed. Luis Eslava, Michael Fakhri, and Vasuki Nesiah (Cambridge: Cambridge University Press, 2017), 3–32, https://www.cambridge.org/core/books/abs/bandung-global-history-and-international-law/spirit-of-bandung/5C0F2DC4165E532553D31B0F7C4F2095.

117 *centered at Merritt College*: Martha Biondi, *The Black Revolution on Campus* (Berkeley: University of California Press, 2012), 41.

117 *a popular icon for Black resistance at the time*: The Panther icon was already in use in Lowndes County, Alabama, as related, for example, in Hasan Kwame Jeffries, *Bloody Lowndes: Civil Rights and Black Power in Alabama's Black Belt* (New York: New York University Press, 2009).

117 *"Abandoning the countryside"*: Frantz Fanon, *The Wretched of the Earth* (New York: Grove Atlantic, 2007), 66.

118 *besmirching the reputation of the city*: For a full rundown, see Floyd Hunter, *Housing Discrimination in Oakland California. A Study Prepared for the Mayor's Committee on Full Employment and the Council of Social Planning of Alameda County* (Berkeley, 1963–64).

118 *"A rat done bit my sister Nell"*: Alexis C. Madrigal, "Gil Scott-Heron's Poem 'Whitey on the Moon,'" *The Atlantic*, May 28, 2011, https://www.theatlantic.com/technology/archive/2011/05/gil-scott-herons-poem-whitey-on-the-moon/239622/.

118 *the environmental devastation they'd wrought*: Nathan McClintock, "From Industrial Garden to Food Desert: Demarcated Devaluation in the Flatlands of Oakland, California," in *Cultivating Food Justice: Race, Class, and Sustainability*, ed. Alison Hope Alkon and Julian Agyeman (Cambridge, MA: MIT Press, 2011), 89–120.

118 *"maximum feasible participation"*: John F. Keilch and Edward M Kirshner, *Money Doesn't Talk, It Swears: The Saga of Radical Economic Development Proposals in Oakland* (1976).

119 *documenting police violence for decades by the 1960s*: In 1951, the East Bay Civil Rights Congress released a major report on police brutality in Oakland. I digitized the document, and it is available here: https://howthebaywasbuilt.com/wp-content/uploads/2019/01/1949-east-bay-civil-rights-congress_police-brutality-report.pdf.

119 *"We would rather die fighting"*: The Spokesman 2, no. 3, 4.

119 *relative to people in white neighborhoods*: Alameda County Public Health Department, *Life and Death from Unnatural Causes: Health and Social Inequity in Alameda County* (Oakland, 2008), https://acphd-web-media.s3-us-west-2.amazonaws.com/media/data-reports/social-health-equity/docs/unnatcs2008.pdf.

119 *"Black people are the children of violence"*: The Spokesman 1, no. 24, 5.

119 *"But she's having a baby"*: Rick Pearlstein, "A Message from Watts," *Newsweek*, November 4, 2008, https://www.newsweek.com/rick-perlstein-new-message-watts-84855.

119 *"The killing of Leonard Deadwyler"*: Thomas Pynchon, "A Journey into the Mind of Watts," *New York Times*, June 12, 1966, https://archive.nytimes.com/www.nytimes.com/books/97/05/18/reviews/pynchon-watts.html?_.

120 *"The colonized masses"*: Fanon, *The Wretched of the Earth*, 33.

120 *Bobby Seale lived with his parents*: Matt O'Brien, "Black Panther Birthplace Flipped and Sold as Trendy Oakland Showpiece," *Mercury News*, March 22, 2012, https://www.mercurynews.com/2012/03/22/black-panther-birthplace-flipped-and-sold-as-trendy-oakland-showpiece/.

120 *not the ghetto but a borderland*: To help see the world of Black Oakland at the time, I pulled all the advertisers from a local magazine from that fall, *The Combination*, and plotted them onto a map, which is available online: https://www.google.com/maps/d/u/0/edit?mid=1FitrgLz6OhPmwLdt21GKlV0jO8m3yX6c&usp=sharing.

121 *it attracted Seale's friend*: Donna Murch, "The Campus and the Street," in *The New Black History: Revisiting the Second Reconstruction*, ed. Manning Marable and Elizabeth Kai Hinton (New York: Palgrave Macmillan, 2011), 53–66.

121 *more direct action*: Sean L. Malloy, *Out of Oakland: Black Panther Party Internationalism During the Cold War* (Ithaca: Cornell University Press, 2017), chapter 2.

121 *see the nation's power in new ways*: Joshua Bloom and Waldo E. Martin, Jr., *Black Against Empire: The History and Politics of the Black Panther Party* (Berkeley: University of California Press, 2016).

122 *"The Black Panther Party"*: Ibid., 270.

122 *"The student strike at SF State"*: Ibid., 274.

122 *"You had to be there"*: "Introduction to the Asian American Movement 1968," Asian American Movement 1968, blogspot.com, January 20, 2008, http://aam1968.blogspot.com/2008/01/introduction-to-asian-american-movement.html.

123 *"The Black Panther Party was never nationalist"*: Ericka Huggins, "An Oral History with Ericka Huggins," interview by Fiona Thomspon, 2007 (Berkeley: Regional Oral History Office, University of California, 2010), https://revolution.berkeley.edu/assets/Oral-History-of-Ericka-Huggins-excerpt-81-103.pdf.

123 *"Freedom is a state"*: Cited in Bloom and Martin, *Black Against Empire*, 275.

123 *"small number of Filipinos"*: Filipino American students, "Phillipine American Collegiate Endeavor," San Francisco State College Special Collections, https://diva.sfsu.edu/collections/strike/bundles/187941.

123 *"not begin to serve"*: L. Ling-chi Wang, "Chinatown and the Chinese," San Francisco State College Special Collections, https://diva.sfsu.edu/collections/strike/bundles/187935.

123 *"our growing ties"*: Third World Liberation Front, "Mexican American Student Confederation," San Francisco State College Special Collections, https://diva.sfsu.edu/collections/strike/bundles/187996.

123 *"I want you to know"*: Cited in Dunn-Salahuddin, "A Forgotten Community, a Forgotten History," 211–34.

124 *Asian American Political Alliance*: Jay Caspian Kang, *The Loneliest Americans* (New York: Crown, 2021), 47.

124 *more or less*: And by more or less, I mean actually; see Third World Liberation Front Constitution, 1969, CES ARC 2015/1, Location 1:1, twLF, box 1, Folder 1, Ethnic Studies Library, University of California, Berkeley, https://revolution.berkeley.edu/twlf-constitution/?cat=440&subcat=0.

124 *reminded her of the Third World days*: Pamela Tau Lee, "Pamela Tau Lee: Community and Union Organizing, and Environmental Justice in the San Francisco Bay Area, 1967–2000," interview by Carl Wilmsen, 1999 (Berkeley: Regional Oral History Office, University of California, 2003), https://digicoll.lib.berkeley.edu/record/218138?ln=en.

126 *Other trustees were executives*: Research Staff of the Graduate Coordinating Committee, "Know Your Trustees," February 3, 1966, http://www.a3mreunion.org/archive/1965-1966/65-66/files_1965-1966/66_KnowYourTrustees.pdf.

126 *SRI and the Hoover Institution itself*: Hoover's Stephan Possony, for example, was a co-author of *The Technology Strategy*, which called for fighting the Cold War with ever more futuristic weapons, including, eventually, Reagan's failed "Star Wars" plan.

126 *"That this complex"*: David Ransom, "The Stanford Complex" (January 1968), http://www.a3mreunion.org/archive/1967-1968/67-68/files_67-68/67-68_Stanford_Complex.pdf.

127 *"What made things move so fast"*: Franklin interview.

127 *"I'm amazed by the accuracy"*: Pacific Studies Center pamphlet on FMC, http://www.a3mreunion.org/archive/1970-1971/70-71_career/files_70-71_career/70-71Career_FMC.pdf.

127 *"Have they ever made nerve gas?"*: Aaron J. Leonard and Conor A. Gallagher, *Heavy Radicals: The FBI's Secret War on America's Maoists* (Alresford, UK: Zero Books, 2015), chapter 3.

128 *it meant something to refuse it*: Harris, *Palo Alto*, 352.

128 *revolutionary Chicano group Venceremos*: Stanford now maintains a Venceremos collection, with many publications by the organization. See https://exhibits.stanford.edu/stanford-pubs/browse/venceremos; see also Venceremos, *Pamoja Venceremos: Together We Will Win* 2, no. 13 (January 22, 1972).

128 *"Thus Stanford has divided the Peninsula"*: Harris, *Palo Alto*, 352–56.

130 *began to dry up*: Bloom and Martin, *Black Against Empire*, 379–81.

130 *battle over the International Hotel in San Francisco*: Kang, *The Loneliest Americans*, 48–56.

13. THE TECHNOLOGY QUESTION

132 *"Is it possible to imagine U.S. agriculture"*: Michael Hardt and Antonio Negri, *Empire* (Cambridge, MA: Harvard University Press, 2000), 397.

132 *Oakland nightlife*: It's fascinating to consider, as the historian Peter Cole did, why they never made common cause with the ILWU. As it turns out, both Panthers and dockworkers hung out at an old bar called the Lamp Post, but politically, not much came of it.

132 *a penthouse "office"*: Dr. Huey P. Newton Foundation Inc., "Dr. Huey P. Newton Foundation Inc. Collection, 1968–1994," Stanford University Manuscripts Division, Archive Box 2, Folder 5, Stronghold Assets.

133 *he lived in that penthouse*: Tim Findley, "Huey Newton: Twenty-Five Floors from the Street," *Rolling Stone*, August 3, 1972, https://www.rollingstone.com/culture/culture-news/huey-newton-twenty-five-floors-from-the-street-176820/.

133 *"nuclear umbrella"*: Guy J. Pauker et al., "In Search of Self-Reliance: U.S. Security Assistance to the Third World Under the Nixon Doctrine. A Report Prepared for Defense Advanced Research Projects Agency" (Washington, DC: Rand Corporation, 1973), https://apps.dtic.mil/sti/pdfs/AD0775353.pdf.

133 *"undeniable interconnection"*: Huey P. Newton, "The Technology Question," in *The Huey P. Newton Reader*, ed. David Hilliard and Donald Weise (New York: Seven Stories Press, 2002), 256–66.

133 *the evidence that he became violent and abusive*: Huggins, "An Oral History with Ericka Huggins," https://digitalassets.lib.berkeley.edu/roho/ucb/text/huggins_. For example, longtime party member Huggins, still a stalwart in the Oakland community, reported in an oral history that Newton repeatedly raped her. Yet Newton remained an important person to her and other former BPP members. "The thing

about Huey Newton is this: he was a good man too. So how I remember him is for the good things that he did do," she said. "I don't choose to remember him as a horrible person because I feel like we all have demons, and I don't excuse a person for the things that they do, operating from their demons, I don't mean that, but I don't hold anybody in place, because I don't want to be held in place either. We all can transform."

135 "We see very little difference": Hilliard and Weise, eds., The Huey P. Newton Reader, 170.

136 "reactionary intercommunalism": For a summary of intercommunalism, see Judson L. Jeffries, Huey P. Newton: The Radical Theorist (Jackson: University Press of Mississippi, 2002), "The Party Line," 62–82.

136 obsessed with the port: Peter Cole, Dockworker Power: Race and Activism in Durban and the San Francisco Bay Area (Champaign: University of Illinois Press, 2018), chapter 2.

137 part of a new form: Elaine Brown, A Taste of Power: A Black Woman's Story (New York: Anchor, 1994), 375.

137 "Oakland was a city": Ibid., 376.

137 "The technology is so great": Huey P. Newton Archive, Box 26, Folder 10–11, archival conversation between Herman Blake and Bobby Seale.

138 unconsolably terrified: The broader story of these production facilities is told by Aihwa Ong, Spirits of Resistance and Capitalist Discipline: Factory Women in Malaysia (Albany: State University of New York Press, 1987).

138 "In one possession which I witnessed": "Changing Role of S.E. Asian Women," Southeast Asia Chronicle—Pacific Research 9, nos. 5–6 (1979): 16, http://www.cpeo.org/pubs /SEAsianWomen.pdf.

138 "How to Handle Hysterical Factory Workers": P. K. Chew, "How to Handle Hysterical Factory Workers," Journal of Occupational Health and Safety (1978): 8.

138 "They are acts of rebellion": Aihwa Ong, "The Production of Possession: Spirits and the Multinational Corporation in Malaysia," in The Anthropology of Organisations, ed. Alberto Corsín Jiménez (Abingdon, UK: Routledge, 2017), 107–21.

139 "In California's Silicon Valley": War on Want, Women Working Worldwide: The International Division of Labour in the Electronics, Clothing and Textiles Industries (London: War on Want, 1985), 16, https://waronwant.org/sites/default/files/Women%20Working %20Worldwide%20web.pdf.

139 impoverished, repressed workers: It should come as no surprise that today, two of the five largest high-technology companies—Apple and Amazon—are the undisputed leaders of supply chain management and logistics, respectively.

140 "highways, airport facilities, and deepwater ports": Richard Stubbs, "Geopolitics and the Political Economy of Southeast Asia," International Journal 44, no. 3 (1989): 517–40, https://doi.org/10.1177/002070208904400301.

140 what we'd now call tech: War on Want, Women Working Worldwide, 4.

140 In one documentary: The Global Assembly Line documentary, directed by Lorraine Gray (1986), available on YouTube: https://www.youtube.com/watch?v=Eq1KJxBI2jQ&t =233s.

140 "By the mid-1970s": Clair Brown and Greg Linden, "Offshoring in the Semiconductor Industry: A Historical Perspective," Brookings Trade Forum on Offshoring of White-Collar Work, July 22, 2005, https://citeseerx.ist.psu.edu/document?repid =rep1&type=pdf&doi=f55cfc2c2bba62354c51d965138a317ac6c4aa4b.

140 "on different sections of the same assembly line": Pacific Studies Center, "Silicon Valley: Paradise or Paradox? The Impact of High Technology Industry on Santa Clara" (Mountain View, CA: Pacific Studies Center, 1977), http://www.cpeo.org/pubs /Paradise.PDF.

141   *"assess the feasibility"*: Blackford, *Pathways to the Present*, 72.

141   *"Workers must uphold their dignity"*: Les Levidow, "Women Who Make the Chips," *Women, Work, and Gender Relations in Developing Countries: A Global Perspective*, ed. Parvin Ghorayshi and Claire Bélanger (London: Bloomsbury, 1996), 44.

141   *"The manual dexterity of the oriental female"*: Annette Fuentes and Barbara Ehrenreich, *Women in the Global Factory* (Boston: South End Press, 1983), 16.

141   *"We hire girls because"*: Levidow, "Women Who Make the Chips," 45.

141   *"hoping to bring back joy and glory"*: "Changing Role of S.E. Asian Women," 4.

141   *"freedom to go out"*: Ibid., 13.

142   *"linkages that contributed"*: Saskia Sassen, *The Mobility of Labor and Capital: A Study in International Investment and Labor Flow* (Cambridge: Cambridge University Press, 1988), 16.

142   *more Asians were immigrating*: Jia Lynn Yang, *One Mighty and Irresistible Tide: The Epic Struggle over American Immigration, 1924–1965* (New York: W. W. Norton, 2020).

143   *400 percent*: Sassen, *The Mobility of Labor and Capital*, 64.

143   *over the following decades*: David Pellow and Lisa Sun-Hee Park, *The Silicon Valley of Dreams: Environmental Injustice, Immigrant Workers, and the High-Tech Global Economy* (New York: New York University Press, 2002), 64.

143   *"Just three things I look for"*: Ibid., 87.

143   *"Company A would hire"*: Ibid., 90–91.

144   *the needs of nuclear weapons researchers*: See Paul N. Edwards, *The Closed World: Computers and the Politics of Discourse in Cold War America* (Cambridge, MA: MIT Press, 1997).

144   *the Korean War laid the foundation for Silicon Valley*: See Lécuyer, *Making Silicon Valley*.

144   *people and employers, at scale*: Dan Bouk, *How Our Days Became Numbered: Risk and the Rise of the Statistical Individual* (Chicago: University of Chicago Press, 2015).

144   *a mechanical database of 26 million people*: "IBM Collators," Columbia University Computing History, http://www.columbia.edu/cu/computinghistory/collators.html, updated December 27, 2022.

145   *PC, Apple, and all the rest*: Fred Turner, *From Counterculture to Cyberculture: Stewart Brand, the Whole Earth Network, and the Rise of Digital Utopianism* (Chicago: University of Chicago Press, 2010).

145   *"In the nineteenth century"*: James Gleick, *The Information: A History, a Theory, a Flood* (New York: Knopf Doubleday, 2011).

146   *The touchstone in the genre*: AnnaLee Saxenian, *Regional Advantage: Culture and Competition in Silicon Valley and Route 128* (Cambridge, MA: Harvard University Press, 1996).

146   *new types of freedom and human flourishing*: Stewart Brand, "Spacewar, Fantastic Life and Death Among Computer Bums," excerpt in *Rolling Stone*, December 7, 1972, https://archive.org/details/19721207rollingstoneexcerptspacewararticlev02/mode/2up.

146   *computing did not end up freeing the people*: For example, see Harry Cleaver's "The Zapatista Effect: The Internet and the Rise of an Alternative Political Fabric," August 10, 2005, https://libcom.org/article/zapatista-effect-internet-and-rise-alternative-political-fabric.

147   *"Silicon Valley: Paradise or Paradox?"*: Pacific Studies Center, "Silicon Valley: Paradise or Paradox?," http://www.cpeo.org/pubs/Paradise.PDF.

148   *immigrants and people of color*: Lisa Sun-Hee Park and David N. Pellow, "Racial Formation, Environmental Racism, and the Emergence of Silicon Valley," *Ethnicities* 4, no. 3 (2004): 403–24.

148 *toxic volatile organic compounds were removed*: Fairchild Semiconductor Corp. (South San Jose Plant), San Jose, CA, https://cumulis.epa.gov/supercpad/SiteProfiles/index .cfm?fuseaction=second.cleanup&id=0901685.

14. TRAUMA

149 *Wilson was one of the few real links*: Self, *American Babylon*, see part 3.

150 *"Proposition 13 was white folks' message to us"*: Ibid., 316.

150 *"Not only did Proposition 13 help to accelerate"*: Ibid., 235.

15. THE SHADOW

157 *the podcast* East Bay Yesterday: "'I'll Die If I Let Go': After the Earthquake, West Oakland Came to the Rescue," February 2018, in *East Bay Yesterday*, produced by *East Bay Yesterday* and Snap Judgment, podcast, 35:14.

162 *Citizens Emergency Relief Team*: Cecily Burt, "West Oakland Citizens Band Together to Fight Freeway After Loma Prieta," *East Bay Times*, October 12, 2009, https:// www.eastbaytimes.com/2009/10/12/west-oakland-citizens-band-together-to-fight -freeway-after-loma-prieta/.

162 *pass resolutions against rebuilding the freeway*: This was, specifically, at Margaret's seventy-third birthday party.

162 *"Freeways have caused high cancer rates"*: Jane Holz Kay, *Asphalt Nation: How the Automobile Took Over America and How We Can Take It Back* (Berkeley: University of California Press, 1998), 48.

162 *"When you say environment"*: Penn Loh and Chappell Hayes, "Freeways, Communities, and Environmental Justice: Oakland's Clean Air Alternative Coalition Fights Environmental Racism: An Interview with Eco-justice Hero Chappell Hayes," *Population & Immigration* 4, no. 2 (1993): 41–42, https://www.jstor.org/stable/4155 4125.

163 *the reunification of West Oakland*: Burt, "West Oakland Citizens."

163 *"If they had tried to rebuild it"*: Ibid.

163 *"October 1989 the earth shook"*: Coalition for West Oakland Revitalization, *West Oakland Visions and Strategies, Phase I Report—"A Community Plan"* (May 31, 1994), https:// cao-94612.s3.amazonaws.com/documents/oak030540.pdf.

164 *third-party enforcer of environmental law*: Michael R. Lozeau, "Tailoring Citizen Enforcement to an Expanding Clean Water Act: The San Francisco Baykeeper Model," *Golden Gate University Law Review* 28, no. 3 (1998): 429–88.

165 *they tied up dredging for years*: Christopher B. Busch, David L. Kirp, and Daniel F. Schoenholz, "Taming Adversarial Legalism: The Port of Oakland's Dredging Saga Revisited," *NYU Journal of Legislation & Public Policy* 2, no. 2 (1999): 179.

165 *fight a toxic waste dump*: United Church of Christ, "A Movement Is Born: Racial Justice and the UCC," https://www.ucc.org/what-we-do/justice-local-church-ministries /efam/environmental-justice/a_movement_is_born_environmental_justice_and _the_ucc/.

167 *the concerns of white, wealthier people*: Jedediah Purdy, *This Land Is Our Land: The Struggle for a New Commonwealth* (Princeton: Princeton University Press, 2021), 102–40.

167 *The new Environmental Protection Agency*: Task Force on the Environmental Problems of the Inner City, "Our Urban Environment and Our Most Endangered People," September 1971, https://nepis.epa.gov/Exe/tiff2png.cgi/2000UW01.PNG?-r +75+-g+7+D%3A%5CZYFILES%5CINDEX%20DATA%5C70THRU75%5CTIFF %5C00000535%5C2000UW01.TIF.

167 *Efforts succeeded at preserving open space*: See maps in Maria J. Santos, James H. Thorne, Jon Christensen, and Zephyr Frank, "An Historical Land Conservation

Analysis in the San Francisco Bay Area, USA: 1850–2010," *Landscape and Urban Planning* 127 (2014): 114–23.

167 *"Particular kinds of open spaces"*: Jason A. Heppler, *Silicon Valley and the Environmental Inequalities of High-Tech Urbanism* (Norman: University of Oklahoma Press, 2024), 70.

168 *"It was so dynamic and real"*: "Founding APEN," https://vimeo.com/364408302.

169 *They protested the fossil fuel giant Chevron*: Ibid.

169 *"an age of fracture"*: Daniel T. Rodgers, *Age of Fracture* (Cambridge, MA: Harvard University Press, 2011).

16. PREDATORY INCLUSION

170 *As a shelf full of books can attest*: Among many, see David M. P. Freund, *Colored Property: State Policy and White Racial Politics in Suburban America* (Chicago: University of Chicago Press, 2010).

171 *close to 63 percent twenty years later*: PK, "Historical Homeownership Rate in the United States, 1890–Present," DQYDJ, https://dqydj.com/historical-homeownership-rate-united-states/.

171 *Late twentieth-century urban home ownership*: Keeanga-Yamahtta Taylor, *Race for Profit: How Banks and the Real Estate Industry Undermined Black Homeownership* (Chapel Hill: University of North Carolina Press, 2019).

172 *simply denied credit*: Martha Poon, "From New Deal Institutions to Capital Markets: Commercial Consumer Risk Scores and the Making of Subprime Mortgage Finance," *Accounting, Organizations and Society* 34, no. 5 (2009): 654–74.

172 *The people buying CDOs often had no idea*: For a detailed analysis of one mortgage-backed security, and a whole bunch of the communications and computation involved, see this microsite from the Financial Crisis Inquiry Commission, "The Story of a Mortgage Security: Inside CMLTI 2006-NC2," http://fcic.law.stanford.edu/resource/staff-data-projects/story-of-a-security.

172 *this did work to a limited extent*: Though this remains contentious. See, for example, Michael S. Barr, "Credit Where It Counts: The Community Reinvestment Act and Its Critics," *NYU Law Review* 80 (2005): 513.

173 *In 1990, in West Oakland*: As described in the Note on Sources, I reconstructed the long HMDA dataset for Oakland and used it to look at long-run capital coming into the neighborhood.

174 *The era of deregulation*: This larger story is told in Herman M. Schwartz, *Subprime Nation: American Power, Global Capital, and the Housing Bubble* (Ithaca: Cornell University Press, 2011).

174 *inflate their profits*: G. R. Krippner, "The Financialization of the American Economy," *Socio-Economic Review* 3, no. 2 (2005): 173–208.

174 *The leveraged buyout*: G. F. Davis, *Managed by the Markets: How Finance Re-shaped America* (Oxford: Oxford University Press, 2011).

175 *a substantial improvement from the 1960s*: U.S. Census Bureau, "Homeownership Rates by Race and Ethnicity: Black Alone in the United States [BOAAAHORUSQ156N]," FRED, Federal Reserve Bank of St. Louis, https://fred.stlouisfed.org/series/BOAAAHORUSQ156N, updated June 7, 2024.

175 *There were 499 foreclosures*: Todd Harvey et al., "Gentrification and West Oakland: Causes, Effects and Best Practices," submitted to the Department of City and Regional Planning in Berkeley, California. University of California at Berkeley, Fall 1999.

175 *"reverse redlining"*: Linda E. Fisher, "Target Marketing of Subprime Loans: Racialized Consumer Fraud & Reverse Redlining," *Harvard Journal of Law & Public Policy* 18, no. 1 (2009): 121.

175   *"resegregation"*: Alex Schafran, *The Road to Resegregation: Northern California and the Failure of Politics* (Berkeley: University of California Press, 2018).

17. "FREE" LAND AND "FREE" MONEY

176   *The basics of closing down the Oakland Army Base*: Much of this history was recorded in an oral history series by Berkeley researchers: Bancroft Library, "Oakland Army Base: About the Project," https://www.lib.berkeley.edu/visit/bancroft/oral-history -center/projects/oakland-army-base.

177   *spilled all over the base for decades*: See ProPublica's data sheet on the cleanup, "Oakland Army Base, Oakland, CA," Bombs in Our Backyard series, *ProPublica*, https:// projects.propublica.org/bombs/installation/CA9213520661002100.

177   *training and placement for West Oakland*: Partnership for Working Families, "Rebuilding the Base," June 2008, https://www.dropbox.com/scl/fi/r3wsnjyw31xw3m13jg3fa/2008 .06_Rebuilding_the_Base-econ-impact-report.pdf?rlkey=j4ipae9ldudusyevad9ppmsf4 &dl=0.

177   *transitional housing*: Aliza Gallo, "The Final Reuse Plan for Oakland Army Base: Letter from the Executive Director," November 18, 2002, https://www.dropbox.com /scl/fi/085066mxoohd7c9kw4h5i/2002-Final-Reuse-Plan-for-Oakland-Army-Base .PDF?rlkey=2n7rsct4hg6xyk9ixrcpmbzl2&dl=0.

177   *a model of inclusive community planning*: Partnership for Working Families, "Rebuilding the Base."

177   *"At the end of the day"*: Margaret Gordon, "Oakland Army Base Oral History Project," interview by Lisa Rubens, 2009 (Regional Oral History Office, Bancroft Library, University of California Berkeley, 2009).

179   *"West Oakland as a community"*: Manuel Pastor, Jr., Chris Benner, Martha Matsuoka, Paul Ong, and Anastasia Loukaitou-Sideris, "The Regional Nexus: The Promise and Risk of Community-Based Approaches to Metropolitan Equity," in *Jobs and Economic Development in Minority Communities*, ed. Paul M. Ong and Anastasia Loukaitou-Sideris (Philadelphia: Temple University Press, 2006), 63–87.

181   *"contentious behavior"*: Teamworks, "7th Street McClymonds Corridor Neighborhood Improvement Initiative—Final Report: Planning Through Year Three Implementation, 1998–2003, for the William and Flora Hewlett Foundation," available here: http://sfteamworks.com/hewlett03.pdf.

181   *the Hewlett Foundation donated $400 million*: "Stanford Receives $400 Million from Hewlett Foundation," *Philanthropy News Digest*, May 1, 2001, https://philanthropynewsdigest .org/news/stanford-receives-400-million-from-hewlett-foundation.

181–82   Neighborhood Knowledge for Change: Jeremy Hays and Steve Costa, *Neighborhood Knowledge for Change: The West Oakland Environmental Indicators Project* (Oakland: Pacific Institute for Studies in Development, Environment and Security, 2002), https://pacinst.org/wp-content/uploads/2013/02/neighborhood_knowledge_for _change3.pdf.

182   *"community-based participatory research"*: Nina Wallerstein and Bonnie Duran, "Community-Based Participatory Research Contributions to Intervention Research: The Intersection of Science and Practice to Improve Health Equity," *American Journal of Public Health* 100, suppl. 1 (2010): S40–S46, doi: 10.2105/AJPH.2009 .184036.

182   *"Oakland sees"*: "Light Shines Amid Oakland Shadows: Downtown Booms, Art Scene Prospers, Despite Crime, Poverty," *Sacramento Bee*, August 11, 1989.

183   *office space in downtown Oakland*: "Leasing Volume Continues Its Upward Trajectory" (CBRE Q2 market report, 2024), https://mktgdocs.cbre.com/2299/7850e40f -b929-4c87-ab36-a94e96c3e3f1-448844325/v032024/oakland-office-figures-q2-2024 .pdf.

183 *"It's so frustrating"*: Vicki Haddon, "In Defense of Oakland—There's More There There Than You Think," *San Francisco Chronicle*, February 2, 2003, D3.

184 *cut him in on deals*: Ryan Tate, "Phil Tagami, Behind-the-Scenes Powerbroker," *SF Business Times*, November 20, 2005, https://www.bizjournals.com/sanfrancisco /stories/2005/11/21/focus4.html.

184 *a $12 million loan from the city*: Ibid.

18. LIFE, DEATH, AND THE PORT

186 *a documentary of the same name*: Unnatural Causes: Is Inequality Making Us Sick? documentary series (broadcast 2008), https://unnaturalcauses.org.

186 *rapidly since 1980*: Alameda County Public Health Department, "Life and Death from Unnatural Causes," https://acphd-web-media.s3-us-west-2.amazonaws.com/media /data-reports/social-health-equity/docs/unnatcs2008.pdf.

186 *people call it "weathering"*: Allana T. Forde, Danielle M. Crookes, Shakira F. Suglia, and Ryan T. Demmer, "The Weathering Hypothesis as an Explanation for Racial Disparities in Health: A Systematic Review," *Annals of Epidemiology* 33 (2019): 1–18.

188 *"To me, that was the point"*: Meanwhile, the epidemiologist and CDC came around. The CDC created a program called the U.S. Small Area Life Expectancy Estimates Project, which has calculated life expectancies for about 90 percent of U.S. census tracts. "After they gave me all that shit, now they're doing it," Iton said. "And I'm like, thank you."

189 *they worked an average of eleven hours per day*: East Bay Alliance for a Sustainable Economy, "Taking the Low Road: How Independent Contracting at the Port of Oakland Endangers Public Health, Truck Drivers, & Economic Growth," September 2007, https://www.dropbox.com/scl/fi/7cyulu4lc37r48pygpdaf/EBASE-Report-Taking -the-Low-Road.pdf?rlkey=z9gxash8dyhi4o8s5fdfkrksp&dl=0.

189 *The union was decimated*: Frank Swoboda, "Teamsters Traveling a Rough, Narrow Road: Trucking," *Los Angeles Times*, April 8, 1994, https://www.latimes.com /archives/la-xpm-1994-04-08-fi-43679-story.html.

189 *"Risk is being shifted"*: Scott L. Cumming, *Blue and Green: The Drive for Justice at America's Port* (Cambridge, MA: MIT Press, 2018), 4.

189 *"We came to America for a better life"*: East Bay Alliance for a Sustainable Economy, "Taking the Low Road," 11.

190 *"It is abundantly clear"*: Ibid., 4.

191 *a Coalition for Clean and Safe Ports event*: "Drivers, Neighbors Push Clean Ports, Decent Jobs," *People's World*, July 6, 2007, https://www.peoplesworld.org/article /drivers-neighbors-push-clean-ports-decent-jobs/.

191 *One of those guys was Bill Aboudi*: Robert Gammon, "City Contractor Faces Labor and Environmental Charges," *East Bay Express*, March 6, 2013, https://www .eastbayexpress.com/oakland/city-contractor-faces-labor-and-environmental -charges/Content?oid=3480996.

192 *the Teamsters' port-organizing model*: James Rufus Koren, "Court Says Port Truckers Can Maintain Independence," *Los Angeles Business Journal*, September 26, 2011, https:// labusinessjournal.com/news/2011/sep/26/court-says-port-truckers-can-maintain -independence/.

193 *"I'll be bringing the experience"*: J. Douglas Allen-Taylor, "Port Commission Nominee a Test for Dellums' Strategies," *Berkeley Daily Planet*, September 28, 2007, https:// www.berkeleydailyplanet.com/issue/2007-09-28/article/28102?headline=Port -Commission-Nominee-a-Test-of-Dellums-Strategies--By-J.-Douglas-Allen-Taylor.

193 *By the mid-2000s, the port*: Jennifer Warburg, "Big Challenges Ahead for the Port of Oakland," *The Urbanist*, June 9, 2015, https://www.spur.org/publications/urbanist -article/2015-06-09/big-challenges-ahead-port-oakland.

195 *Maritime Air Quality Improvement Plan*: Port of Oakland, "Maritime Air Quality," https://www.portofoakland.com/community/environmental-stewardship/maritime-air-quality-improvement-plan/.

196 *emissions from trucks in West Oakland by 2015*: Port of Oakland, "Truck Diesel Emissions at Port of Oakland Down 98 Percent in Past Decade," Port of Oakland press release, October 28, 2016, https://www.portofoakland.com/press-releases/truck-diesel-emissions-port-oakland-98-percent-past-decade/.

19. THE REVOLUTION CAFE

197 *So much Black wealth got wiped out*: Heather Long and Andrew Van Dam, "The Black-White Economic Divide Is as Wide as It Was in 1968," *Washington Post*, June 4, 2020, https://www.washingtonpost.com/business/2020/06/04/economic-divide-black-households/.

198 *more than 8,500 percent*: This is from a dataset that I reconstructed of HMDA data going as far back as possible.

199 *"The first collateralized mortgage obligation"*: Michael Osinski, "My Manhattan Project," *New York*, March 27, 2009, https://nymag.com/news/business/55687/.

199 *"a time of innovation"*: Poon, "From New Deal Institutions to Capital Markets."

200 *ballooned for the next several years*: Danielle DiMartino and John V. Duca, "The Rise and Fall of Subprime Mortgages," *Economic Letter: Insights from the Federal Reserve Bank of Dallas* 2, no. 11 (November 2007), https://fraser.stlouisfed.org/title/economic-letter-6362/rise-fall-subprime-mortgages-607610.

200 *a single payment*: "The Giant Pool of Money," in *This American Life*, produced by WBEZ Chicago, podcast, May 9, 2008, 58:51, https://www.thisamericanlife.org/355/the-giant-pool-of-money.

201 *the most profitable path for the banks*: See, for example, the devastating analysis in Aaron Glantz, *Homewreckers: How a Gang of Wall Street Kingpins, Hedge Fund Magnates, Crooked Banks, and Vulture Capitalists Suckered Millions Out of Their Homes and Demolished the American Dream* (New York: Custom House, 2019).

201 *One Urban Strategies Council analysis*: Urban Strategies Council, "Who Owns Your Neighborhood? The Role of Investors in Post-Foreclosure Oakland," June 2012, https://ebho.org/wp-content/uploads/2011/09/WhoOwnsYourNeighborhood_ExecSum.pdf.

202 *the industrial suburbs*: Schafran, *The Road to Resegregation*.

202 *a market value of more than $20 billion*: Darwin BondGraham, "The Rise of the New Land Lords," *East Bay Express*, February 12, 2014, https://www.eastbayexpress.com/oakland/the-rise-of-the-new-land-lords/Content?oid=3836329&showFullText=true.

202 *"We are gratified"*: GI Partners, "GI Partners Sells Its Waypoint Portfolio to Colony Starwood," June 5, 2017, https://www.gipartners.com/news/gi-partners-sells-its-waypoint-portfolio-to-colony-starwood.

202 *eviction notices to residents*: "Neil Sullivan / REO Homes," Anti-Eviction Mapping Project, https://www.antievictionmap.com/neill-sullivanreo-homes-1.

202 *along with his friends*: Darwin BondGraham, "Neil Sullivan's Oakland," *East Bay Express*, April 2, 2014, https://www.eastbayexpress.com/oakland/neill-sullivans-oakland/Content?oid=3879812.

202 *by the end of 2021*: Kevin Schaul and Jonathan O'Connell, "Investors Bought a Record Share of Homes in 2021. See Where," *Washington Post*, February 16, 2022, https://www.washingtonpost.com/business/interactive/2022/housing-market-investors/.

202 *rents went up 80 percent*: Julia Prodis Sulek and Kaitlyn Bartley, "House Poor: How Price Hikes Hurt the Most Vulnerable," *Mercury News*, December 8, 2019, https://extras.mercurynews.com/pricewepay/part2/.

203 *in the tech industry*: Alexis C. Madrigal, "Who's Really Buying Property in San Francisco?," *The Atlantic*, April 19, 2019, https://www.theatlantic.com/technology /archive/2019/04/san-francisco-city-apps-built-or-destroyed/587389/.

204 *the real big money*: For example, "Preqin Special Report: The Private Equity Top 100, February 2017," https://docs.preqin.com/reports/Preqin-Special-Report-The -Private-Equity-Top-100-February-2017.pdf.

205 *more than $113 trillion under management*: Thinking Ahead Institute, "Top 500 Investment Managers See Assets Drop by $18 Trillion," press release, October 23, 2023, https://www.thinkingaheadinstitute.org/news/article/top-500-investment -managers-see-assets-drop-by-18-trillion/.

205 *bought and sold like a commodity*: Mathias Hoffmann and Iryna Stewen, "Holes in the Dike: The Global Savings Glut, U.S. House Prices and the Long Shadow of Banking Deregulation," CESifo Working Paper No. 5332, Center for Economic Studies and Ifo Institute (CESifo), Munich, 2015, https://www.econstor.eu/bitstream/10419 /110847/1/cesifo_wp5332.pdf.

205 *through the Japanese SoftBank*: Theodore Schleifer, "SoftBank, the Most Powerful— and Controversial—Tech Investor in Silicon Valley," Vox.com, May 10, 2019, https:// www.vox.com/recode/2019/5/10/18563267/softbank-vision-fund-explainer-uber -wework-slack-ipo.

205 *another plane of economic existence*: Pacific Maritime Association, "2018 Annual Report," www.pmanet.org/wp-content/uploads/2019/04/2018-PMA-Annual-Report .pdf.

20. THE BASE

207 *downtown Oakland's resurgence*: Jim Harrington, "Oakland's Fox Theater Ready to Reopen Friday," *East Bay Times*, February 4, 2009, https://www.eastbaytimes.com /2009/02/04/oaklands-fox-theater-ready-to-reopen-friday/.

209 *"behind the scenes powerbroker"*: For the tangled, dense web of Oakland politics, see, for example, Chip Johnson, "Building Power in Oakland / Interesting Parallels Between Developers and Politicians," *SFGate*, November 22, 2004, https://www.sfgate .com/bayarea/johnson/article/Building-power-in-Oakland-Interesting-parallels -2670117.php.

209 *"We are pursuing a new economy"*: Blanca Torres, "Former Oakland Army Base Kicks Off $500 Million First Phase," *San Francisco Business Times*, October 31, 2013, https:// www.bizjournals.com/sanfrancisco/blog/real-estate/2013/10/oakland-army-base -construction-starts.html.

210 *"An unprecedented coalition"*: John Barna wrote an op-ed that Phil Tagami had in his files called "Promises Made, Promises Unmet." Unpublished, I believe: https:// www.dropbox.com/scl/fi/kj9ich46r46391lyootdc/Barna-Promises-Made-7219.pdf ?rlkey=91ro2b6g1yjyj6tzz9jizkig7&dl=0.

210 *1.2 billion square feet of warehouses*: "Prologis, Inc. REIT Profile," reitnotes.com, updated June 15, 2024, https://www.reitnotes.com/reit/symbol/PLD.

210 *Amazon its largest customer*: Patrick Clark and Matthew Boyle, "Two Companies Are Dominating the Battle for Warehouse Space," Bloomberg, October 29, 2019, https://www.bloomberg.com/news/articles/2019-10-29/blackstone-prologis-pull -away-in-amazon-fueled-warehouse-battle.

210 *Over the next seven years*: These numbers are pulled together from Amazon's annual SEC filings.

211 *The company likes to say*: Prologis, "Oxford Economics: Prologis Properties Facilitate Goods Equal to Nearly 3% of the World's GDP and House 1.1. Million Jobs," December 14, 2022, https://www.prologis.com/news-research/press-releases/oxford -economics-prologis-properties-facilitate-goods-equal-nearly-3.

212 *"signal[ing] the incoherence"*: Chris Fan, "Jean Quan and the Death of Asian America," *Hyphen*, November 21, 2011, https://hyphenmagazine.com/blog/2011/11/jean-quan -and-death-asian-america.

213 *"In the ILWU"*: All city council meeting videos can be viewed at the Oakland City Government website, https://www.oaklandca.gov/services/view-city-council-meeting -schedule-agendas-minutes-and-video.

214 *"There are also rumblings"*: Phillip Matier and Andrew Ross, "2 on Oakland Council Going Head to Head," *San Francisco Chronicle*, July 25, 2012, C1.

214 *Her last stand*: The full city council meeting video is available; see City of Oakland, "Special Oakland City Council," July 26, 2012, https://oakland.granicus.com/player /clip/1097.

21. BIG SHIT

219 *"It's been said"*: A video of the groundbreaking has been uploaded to YouTube: Architectural Dimensions, "Oakland Army Base Groundbreaking Project," November 1, 2013, 0:7:11, https://www.youtube.com/watch?v=w9YqZlamlD8.

221 *seemed to agree it would be "stupid"*: Pete Cashman's deposition: United States District Court Northern District of California San Francisco Division, *Oakland Bulk & Oversized Terminal LLC v. City of Oakland*, videotaped deposition of Patrick J. Cashman, San Francisco, CA, August 28, 2017, https://www.dropbox.com/scl /fi/b3ezn78j0xqk611pxnhhn/18_Cashman-Deposition-Full-Transcript.pdf?rlkey =4p17odb6af65p5cvufwyl7fkv&dl=0.

221 *"One bulk material"*: Anti-coal activists have archived a copy of the newsletter here: Oakland Global Trade & Logistics Center & California Capital & Investment Group, "Oakland Global News, 2013," issue 4, https://nocoalinoakland.info/wp -content/uploads/2023/04/OaklandGlobalNewsletter_Issue-4_December2013 .pdf.

222 *"We must accelerate the use of modern renewables!"*: Phillip H. Tagami (@ptagami), "This chart illustrates our challenge to save the planet. We must accelerate the use of modern renewables!," Twitter, February 1, 2014, 6:45 p.m., https://twitter.com /ptagami/status/429807677460742145?s=20.

222 *"staff infers"*: Oakland Board of Port Commissioners, "Agenda, Meeting of the Board of Port Commissioners, February 27, 2014, 1:00 p.m.," https://www.sierraclub.org /sites/www.sierraclub.org/files/View%20Report%20(3).pdf.

222 *dozens of new power stations*: "Japan to Stop Building Unabated Coal Power Plants, PM Kishida Tells COP28," Reuters, December 1, 2023, https://www.reuters.com /world/asia-pacific/japan-stop-building-unabated-coal-power-plants-pm-kishida -tells-cop28-2023-12-01/.

225 *"Fight the bad and build the new"*: Mari Rose Taruc, "California Environmental Justice Organizers Are Writing the Future of Climate Plans," *Nonprofit Quarterly*, December 22, 2023, https://nonprofitquarterly.org/california-environmental-justice -organizers-are-writing-the-future-of-climate-plans/.

22. KEEP IT IN THE GROUND

229 *"Project Could Transform Local Coal Market"*: *The Reaper* took down the article, but it can be found here: http://www.riversimulator.org/farcountry/Sitla/Exchange /ProjectCouldTransformLocalCoalMarketToInternational.pdf.

230 *"We've had an unfortunate article appear"*: Activists cite Holt's email repeatedly: No Coal in Oakland, "Utah Passes Bill to Invest $53 Million in Oakland Coal Terminal," March 11, 2016, https://nocoalinoakland.info/utah-legislature-votes-53-million-to -invest-in-oakland-coal-terminal/.

231 *got behind the terminal*: Darwin BondGraham, "Buying Support for Coal," *East Bay*

Express, September 21, 2015, https://www.eastbayexpress.com/SevenDays/archives /2015/09/21/buying-support-for-coal.

232 *not a single coal car*: Ibid.

233 *drop by more than 60 percent*: EIA, "Table 3.1.A, Net Generation by Energy Source: Total (All Sectors), 2012–2022," https://www.eia.gov/electricity/annual/html/epa_03_01_a .html.

233 *"The rough justice of the market reaction"*: Tom Sanzillo, "IEEFA Update: The Coal Deal That Collapsed to Nobody's Surprise," Institute for Energy Economics and Financial Analysis, November 17, 2017, https://ieefa.org/ieefa-update-coal-deal-collapsed -almost-overnight/.

233 *what they say is 20 percent*: Alex Katz, "Continuing the Fight Against Coal in Oakland," UC Berkeley International & Executive Programs, https://iep.berkeley.edu /content/continuing-fight-against-coal-oakland.

233 *more than 8 billion tons*: IEA, "Global Coal Demand Set to Remain at Record Levels in 2023," July 27, 2023, https://www.iea.org/news/global-coal-demand-set-to-remain -at-record-levels-in-2023.

233 *"We applied for and the City approved"*: Darwin BondGraham, "Coal Terminal Developer Suing City of Oakland, Seeking to Overturn Ban on Fossil Fuel Exports," *East Bay Express*, December 8, 2016, https://www.eastbayexpress.com/SevenDays /archives/2016/12/08/coal-terminal-developer-suing-city-of-oakland-seeking-to -overturn-ban-on-fossil-fuel-exports.

235 *a huge settlement*: Maria Dinzeo, "Ninth Circuit Posed to Uphold Jury Verdict in Fatal Shooting by BART Officer," *Courthouse News Service*, January 13, 2022, https://www .courthousenews.com/ninth-circuit-poised-to-uphold-jury-verdict-in-fatal-shooting -by-bart-officer/.

237 *a voluntary set of risk management guidelines*: Equator Principles, "About the Equator Principles," https://equator-principles.com/about/.

237 *"to demonstrate effective Stakeholder Engagement"*: Equator Principles, "Equator Principles III June 2013," https://equator-principles.com/wp-content/uploads/2017/03 /equator_principles_III.pdf.

241 *"You hardly ever see an opinion"*: Dan Brekke, "Federal Judge Throws Out Oakland's Ban on Coal Shipments Through Planned Terminal," *The California Report*, KQED, May 15, 2018, https://www.kqed.org/news/11668412/federal-judge-throws-out-oak lands-coal-ban.

245 *"We share the same principles"*: Osamu Tsukimori, "Is Japan's 'Clean Coal' Initiative Lagging Behind the Rest of the World?," *Japan Times*, September 24, 2019, https:// www.japantimes.co.jp/news/2019/09/24/business/japan-clean-coal-climate -change-pollution/#.XgG5vxdKhsM.

246 *"Restricting any commodity on political grounds"*: BondGraham, "Coal Terminal Developer Suing City of Oakland."

23. IN THE INTEREST OF OTHERS

250 *single hottest neighborhood real estate market in the country*: Melia Robinson, "Tour the Little-Known California 'Micro-Hood' That's Suddenly the Hottest Housing Market in America," *Business Insider*, June 11, 2017, https://www.businessinsider.com /photos-of-bushrod-oakland-what-its-like-2017-6.

252 *"In general, if rates overall go up"*: Alexis C. Madrigal, "3 Million Uber Drivers Are About to Get a New Boss," *The Atlantic*, April 10, 2018, https://www.theatlantic.com /technology/archive/2018/04/uber-driver-app-revamp/557117/#.

253 *all their gore and grimness*: See Susan Fowler, *Whistle Blower: My Journey to Silicon Valley and Fight for Justice at Uber* (New York: Viking, 2020), and Mike Isaac, *Super Pumped: The Battle for Uber* (New York: W. W. Norton, 2019).

254 *"People don't realize"*: Alexis C. Madrigal, "Free Money at the Edge of the Tech Boom," *The Atlantic*, October 19, 2017, https://www.theatlantic.com/technology /archive/2017/10/stockton_ubi_basic_income/543036/.

257 *civil rights work of the mid-century*: International Longshore & Warehouse Union, "Bill Chester: ILWU Civil Rights and Community Leader, 1938–1969," ILWU Oral History Project, Vol. VI, Part I, May 18, 2004, https://www.ilwu.org/oral-history-of -bill-chester/.

259 *"We demand fair trade"*: International Longshore & Warehouse Union, "ILWU International President Brian McWilliams' Speech at the WTO Labor Rally," November 30, 1999, https://www.ilwu19.com/history/wto/speech.htm.

260 *"to explain why"*: John S. Ahlquist and Margaret Levi, *In the Interest of Others: Organizations and Social Activism* (Princeton: Princeton University Press, 2013).

260 *"Participants adopted a decentralized"*: Luis A. Fernandez, *Policing Dissent: Social Control and the Anti-Globalization Movement* (New Brunswick, NJ: Rutgers University Press, 2008), 6.

260 *"Thanks to the Net"*: Naomi Klein, *No Logo: No Space, No Choice, No Jobs* (New York: Picador, 2009), 453.

261 *"In contrast to imperialism"*: Hardt and Negri, *Empire*, xii.

261 *"The spatial divisions"*: Ibid., xiii.

261 *"As a mass demographic shift"*: Ibid., 297.

265 *such as commercial fishing*: See Ian Urbina, *The Outlaw Ocean: Journeys Across the Last Untamed Frontier* (New York: Vintage, 2019).

266 *L's story was one of hundreds of thousands*: Republic of the Philippines Department of Transportation Maritime Industry Authority, "Marina Statistical Report, 2017–2021," https://marina.gov.ph/wp-content/uploads/2022/06/2017-2021-MARINA-Statistical -Report_FINAL_revised.pdf.

24. TRANSPANDEMIC

269 *"Suddenly, all supply chains seem vulnerable"*: Rosemary Coates, "Coronavirus and Your Global Supply Chain—Part 2—Rising Panic," *Supply Chain Management Review*, February 10, 2020, https://www.scmr.com/article/coronavirus_and_your _global_supply_chains_part_2_rising_panic.

269 *the sharpest decline in its customer sentiment survey ever*: Prologis, "Logistics Activity a Reflection of Covid-19 Disruption," April 28, 2020, https://www.prologis.com /insights/global-insights-research/logistics-activity-reflection-covid-19-disruption.

269 *cleared out the distributors*: Andrew Moore, "How the Coronavirus Created a Toilet Paper Surge," NC State University, *College of Natural Resources News*, May 19, 2020, https://cnr.ncsu.edu/news/2020/05/coronavirus-toilet-paper-shortage/.

270 *United Microelectronics Corp.*: Prableen Bajpai, "An Overview of the Top 5 Semiconductor Foundry Companies," October 1, 2021, https://www.nasdaq.com/articles /an-overview-of-the-top-5-semiconductor-foundry-companies-2021-10-01.

270 *And that's what happened*: Michael Kanellos, "Soaring Costs of Chipmaking Recast Industry," CNETNews, January 23, 2003, https://archive.ph/20130119214941/http:// news.com.com/Semi+survival/2009-1001_3-981418.html.

271 *"Today we see such concentration"*: "Promoting Competition, Growth, and Privacy Protection in the Technology Sector: The Lessons of America's Supply Chain Crisis," Barry C. Lynn, Executive Director Open Markets Institute, Testimony Before the Senate Finance Committee Subcommittee on Fiscal Responsibility and Economic Growth, 117th Congress, December 7, 2021, https://www.finance.senate.gov /download/12072021-lynn-testimony.

271 *rather than reinvest in long-term capacity*: Garphil Julien, "To Fix the Supply Chain Mess, Take On Wall Street," *Washington Monthly*, January 17, 2022, https://

washingtonmonthly.com/2022/01/17/to-fix-the-supply-chain-mess-take-on-wall
-street/.

272   *the lack of chips*: Michael Wayland, "Chip Shortage Expected to Cost Auto Industry
$210 Billion in Revenue in 2021," CNBC, September 23, 2021, https://www.cnbc
.com/2021/09/23/chip-shortage-expected-to-cost-auto-industry-210-billion-in-2021
.html.

272   *"the industry of industries"*: Susan Helper and Evan Soltas, "Why the Pandemic Has
Disrupted Supply Chains," June 17, 2021, The White House, https://www.whitehouse
.gov/cea/written-materials/2021/06/17/why-the-pandemic-has-disrupted-supply
-chains/.

272   *"The weakest link"*: Hassan Khan (@hassankhan), "As journalists plumb further
into supply chain bottlenecks, we're beginning to see a consistent theme: the weakest link in the supply chain is often an inexpensive part. These nodes have been,"
Twitter, November 8, 2021, 8:35 a.m., https://twitter.com/hassankhan/status/145
7748600838975495.

274   *"Cities were like ships"*: Richard White, *The Republic for Which It Stands: The United
States During Reconstruction and the Gilded Age, 1865–1896* (Oxford: Oxford University
Press, 2017), 485.

275   *"servant economy"*: Alexis C. Madrigal, "The Servant Economy," *The Atlantic*, March 6,
2019, https://www.theatlantic.com/technology/archive/2019/03/what-happened
-uber-x-companies/584236/.

277   *highest monthly number ever recorded*: See Hamza Abdelrahman and Luiz Edgard Oli-
veira, "Pandemic Savings Are Gone: What's Next for U.S. Consumers?," *Federal Reserve Bank of San Francisco Blog*, May 3, 2024, https://www.frbsf.org/research-and
-insights/blog/sf-fed-blog/2024/05/03/pandemic-savings-are-gone-whats-next-for-us
-consumers/. See also Justin Lahart, "How Much Savings Do Americans Have Left,
Anyway?" *Wall Street Journal*, September 26, 2023, https://www.wsj.com/economy
/central-banking/how-much-savings-do-americans-have-left-anyway-f1bac32e.

277   *The average shipping time from China*: Jude Abraham, "Shipping & Freight Cost Increases, Current Shipping Issues, and Shipping Container Shortage [2024]," *Freight
101 Blog*, Freightos, February 1, 2024, https://www.freightos.com/freight-resources
/coronavirus-updates/.

277   *almost eighteen days to unload*: "Port of Los Angeles Signal—Shipping Delays,"
January 28, 2022, https://www.dropbox.com/scl/fi/8r8jg3srn0zv1rnw12f8j/Jan
-28-2022-Port-of-LA-Signal-ship-delays.pdf?rlkey=4abi1spvdsn39pp5ma07bmlw3
&dl=0.

278   *$800 per container*: Bill Mongelluzzo, "Container Lines Suffer Brutal Trans-Pacific
Contract Season," *Journal of Commerce*, June 1, 2016, https://www.joc.com/maritime
-news/trade-lanes/trans-pacific/container-lines-suffer-brutal-trans-pacific-contract
-season_20160601.html.

278   *an 8,500 percent increase*: Sujata Rao and Jonathan Saul, "Shipping Costs—Another
Danger for Inflation-Watchers to Navigate," Reuters, December 10, 2021, https://
www.reuters.com/markets/commodities/shipping-costs-another-danger-inflation
-watchers-navigate-2021-12-10/.

278   *erase the huge losses of previous years*: Lauren Etter and Brendan Murray, "Shipping
Companies Had a $150 Billion Year, Economists Warn They're Also Stoking Infla-
tion," Bloomberg, January 17, 2022, https://www.bloomberg.com/news/features
/2022-01-18/supply-chain-crisis-helped-shipping-companies-reap-150-billion-in
-2021.

278   *losing something like $10 billion a year*: Alexis C. Madrigal, "The Ships, the Tugs, and
the Port," *Medium*, March 14, 2017, https://medium.com/containers/episode-3-the
-ships-the-tugs-and-the-port-7825956a7101.

278 *"When they've got product in China"*: "How Supply Chain Backups Threaten to Leave Store Shelves Bare," October 5, 2021, in *KQED Forum with Mina Kim and Alexis Madrigal*, San Francisco, KQED-FM, https://www.kqed.org/forum/2010101885812/how -supply-chain-backups-threaten-to-leave-store-shelves-bare.

279 *just barely crossed 1 million*: Pacific Merchant Shipping Association (PMSA), "December's TEU Numbers (Mind the Gaps)," *PMSA West Coast Trade Report*, January 2022, https://www.pmsaship.com/wp-content/uploads/2022/01/West-Coast-Trade -Report-January-2022.pdf.

280 *"a convoluted patchwork"*: Hugh R. Morley, "JOC Chassis Explainer: Shipping Industry Seeks Solutions," *Journal of Commerce*, July 16, 2018, https://www.joc.com/article /joc-chassis-explainer-shipping-industry-seeks-solutions_20180716.html.

280 *boxes piled up in a storage yard*: "Empty Containers Pile Up at Port of Oakland After Korean Shipper's Bankruptcy," CBS News Bay Area, October 8, 2016, https://www .cbsnews.com/sanfrancisco/news/containers-port-oakland-pile-up/.

280 *It's a pure loss*: Alana Semuels, "The Truck Driver Shortage Doesn't Exist. Saying There Is One Makes Conditions Worse for Drivers," *Time*, November 12, 2021, https:// time.com/6116853/truck-driver-shortage-supply-chain/.

25. THE GRAND BARGAIN

282 *a scathing report about its failings*: California Environmental Justice Alliance, "Lessons from California's Community Emissions Reduction Plan: AB 617's Flawed Implementation Must Not Be Repeated," 2021, https://caleja.org/wp-content/uploads /2021/05/CEJA_AB617_r4-2.pdf.

283 *the capacity of the organization*: ProPublica Nonprofit Explorer, "West Oakland Environmental Indicators Project," https://projects.propublica.org/nonprofits/organizations /202384563, updated May 23, 2024.

283 *Margaret opponent*: Darwin BondGraham, "Oakland's Mayor's Race: Unions, Coal Terminal Developers, and Others Are Spending Big on a Few Candidates," *The Oaklandside*, October 3, 2022, https://oaklandside.org/2022/10/03/oakland-mayor -campaign-2022-unions-coal-terminal-independent-expenditure-committee/.

284 *fleeing the wars of Southeast Asia*: Maanvi Singh, "From Homeless to City Hall: The Hmong American Mayor Making History in Oakland," *The Guardian*, November 24, 2022, https://www.theguardian.com/us-news/2022/nov/24/sheng-thao-hmong -american-mayor-oakland.

285 *found its way to the Wedgewood subsidiary*: Katie Ferrari, "The House on Magnolia Street," *Curbed San Francisco*, April 29, 2020, https://sf.curbed.com/2020/4/29/21240456 /moms-4-housing-oakland-house-history.

289 *the informal settlement:* Erin Baldassari, "The End of Wood Street: Inside the Struggle for Stability, Housing on the Margins of the Bay Area," *KQED: The California Report Magazine*, May 17, 2023, https://www.kqed.org/news/11949327/the-end-of-wood -street-inside-the-struggle-for-stability-housing-on-the-margins-of-the-bay-area.

Acknowledgments

I've been dreading writing these acknowledgments because I've been working on this book for so long that I feel like half of Oakland has helped me out at one point or another. No way out but through, I guess.

I should start with the librarians and archivists, who are the best of us. Dorothy Lazard is thanked elsewhere, in the notes, but she is the doyenne of Oakland history. At the African American Museum and Library, many thanks are due to Sean Dickerson. The rest of the Oakland history nerds deserve special mention, too: Liam O'Donoghue, Moriah Ulinskas, Sue Mark, and Gene Anderson. You all inspire me. And Liam provided crucial feedback when I needed it. My thanks also to Aaron Seltzer, from the National Archives in San Bruno; Gina Bardi, at the San Francisco Maritime National Historical Park Research Center; and Leslie Berlin, then of the Silicon Valley Archives at Stanford. But really: thank you to all the librarians and archivists.

During the course of writing this book, AnnaLee Saxenian helped me become a visiting scholar at the School of Information at UC Berkeley, which was immensely helpful. In the geography department, Brandi Summers provided so many insightful conversations. Carolina Reid provided encouragement and conceptual help with putting together the Home Mortgage Disclosure Act data I assembled for West Oakland. Two Berkeley-trained thinkers also provided me with fortitude and feedback: Rasheed Shabazz, a scholar's scholar, and Alex Werth, a deep and sensitive historian of this place. Jennifer Light's work and her gracious counsel were tremendously useful.

I've been working on this book for so long that I've had four sets of colleagues. It's ridiculous how many people have had to hear me go on about this book. I started this project while working at Univision for Isaac Lee and Daniel Eilemberg and Dodai Stewart with tremendous colleagues Kristen Brown, Hillary Frey, Kashmir Hill, Kevin Roose, Joyce Tang, and many others. I grew up, as a person and a writer, while I was at *The Atlantic*. I can't imagine my writing life without the editors I've had there: Ross Andersen, Adrienne LaFrance, Ellen Cushing, Paul Bisceglio, and Sarah Laskow. Bob Cohn and J. J. Gould have been the key mentors of my career. In a roundabout way, Ta-Nehisi Coates kicked this project off, when we were building out the digital presentation for "The Case for Reparations." I'm in awe of what these colleagues and friends have done: Yoni Appelbaum, Ian Bogost, Kasia Cieplak-Mayr Von Baldegg, Ellen Cushing, James Fallows, Conor Friedersdorf, Megan Garber, Jeffrey Goldberg, David Graham, Adam Harris, Becca Rosen, Swati Sharma, Derek Thompson, Katherine Wells, Ed Yong, and Sarah Zhang.

At KQED, the team is so generous and wonderful that I could fill these whole acknowledgements with thanks for them. I'm so grateful they let me join the powerhouse *Forum* team: Danny Bringer, Susan Britton, Judy Campbell, Susan Davis, Francesca Fenzi, Marlena Jackson-Retondo, Mina Kim, Jennifer Ng, Ariana Proehl, Blanca Torres, Ethan Toven-Lindsey, Caroline Smith, and Grace Won.

All around the station, I've learned so much from our people. Thank you: Olivia Allen-Price, Erin Baldassari, Adhiti Bandlamudi, Madi Bolaños, Pendarvis Harshaw, Marisa Lagos, Guy Marzorati, Lesley McClurg, Gabe Meline, Ezra David Romero, Farida Jhabvala Romero, Scott Shafer, Nastia Voynovskaya, and Brian Watt.

The whole team at the COVID Tracking Project inspired and grounded me during the terrible period of the early pandemic. I still can't believe we did all that. Erin Kissane! All those calls, all you were dealing with. I'd run something with you any day. Special shout-outs to: Artis Curiskis, Amanda French, Alice Goldfarb, Julia Kodysh, Elliott Klug, Michal Mart, Theo Michel, Kevin Miller, Kara Oehler, Ryan Panchadsaram, Prajakta Ranade, Jessica Malaty Rivera, Kara Schechtman, Peter Walker. Honestly, I don't think I've processed any part of that experience except that you all are my heroes.

Different scenes in Oakland have each contributed to my sense of this place. The staff and students at Chapter 510, the youth writing center, have always kept the creativity of this place at the front of my mind. Same for my mentee, Miguel, who I am so damn proud of. The whole Renegade Running crew has provided a shining example of community building, gotten me back into shape, and placed me into fellowship with so many different kinds of people. I want to especially shout out Victor Diaz, who dreamt up Renegade, and Miya Hirabayashi, who was the only person I trusted to take my headshot for this book.

The bookstores (and booksellers!) of the Bay Area are a necessary thank-you, too. I've had the opportunity to do interviews and readings at different places, and I love what our booksellers do. You make me want to be a part of the world of letters. Thank you to the booksellers at Book Passage, City Lights, Clio's, East Bay Booksellers, Green Apple Books, Marcus Books, Pegasus Books, Point Reyes Books, Wolfman Books (RIP), Womb House Books. The good fight goes on.

My friend and former colleague Kyana Moghadam once told me that the Bay Area has such a small creative scene that she imagined everyone out here quietly holding hands. And that has been the case for me. I've learned from and been inspired by so many of you: Mandy Aftel, Reem Assil, Catherine Bracy, Tracy Clark-Flory, Twilight Greenaway, Kate Goldstein-Breyer, Jasmine Guillory, Liz Hernández, Clara Jeffery, Ada Limón, Roman Mars, Courtney Martin, Douglas McGray, Holly Mulder-Wollan, Davia Nelson, Tommy Orange, Jennifer Pahlka, Brontez Purnell, Nikki Silva, Rebecca Skloot, Makshya Tolbert, Carvell Wallace, Jacob Ward, and Malia Wollan.

I owe a little something to so many different bars and cafes: the Punchdown, where I wrote sections and came to the gospel of natural wine thanks to Esteechu and the crew; Prizefighter, where I cooked portions of this book with Robin; Redfield, which is low-key the best place to work; Golden Sardine, which has me wanting to write poetry and drink Riesling; and Tallboy, which reminds me that we can have nice things.

A few people, through their generosity and spirit, helped push this book across the finish line. Emily Bell had lunch with me at just the moment I needed it. Marthine Satris helped rescue me from the depths of my book despair during a chance encounter at Ordinaire. And she hooked me up with Daniela Blei, who swooped in to save my endnotes. George Saunders nudged this project back onto the rails when it was in serious trouble. Jon Mooallem told me that "books are made of other books" while we were pushing kids on swings on Bainbridge Island, and I've used that koan to keep myself going at many different junctures.

Of course, I must thank my editor, Sean McDonald, who really let me run with this project. We did it! It took us ten years, but we did it. Thank you. And a big shout-out to the rest of the team at MCD, the labelmates, and the broader Farrar, Straus and Giroux universe.

It's ridiculous to save family (and chosen family) for the end, but that's the tradition, so here we go.

My neighbors saved my sanity through the pandemic (and before and beyond) and have made the extended Colbyverse a place I cherish: Amanda, Ben, Daniela, Diane, Erika, Hannah, Heather, John, Jonathan, Joyce, Lisa,

Lynne, Matt, Mousa, Neel, Pamela, Rose, Tiffany, the erstwhile Hippo House residents, and Elmer, who will always be the mayor of the block in my heart.

For several years while writing this book, I spent nearly all my working days at the Murray Street Media Lab, working side by side with Robin Sloan. A guy could not ask for a smarter, nicer, more intellectually engaged friend than Robin. I wrote the decisive draft of this book sitting on the floor of his house in Fresno, with him cheering me on. He also introduced me to an incredible reader of this book, Dan Bouk.

Going back even further in time, my friend, my brother, Teddy Wright, has shaped me more than I could possibly recount here. Who knew a chance encounter a few days into our adulthood would become a friendship that's lasted . . . well, all of adulthood.

I don't have the words to express my admiration of and pride about Robinson Meyer. We all saw his talent when he was truly and actually twenty-one years old, and he's far exceeded even those lofty expectations as a writer and a friend.

The Monday dinner crew has been the heartbeat of the last half decade: Cara Rose DeFabio, Thao Nguyen, Samin Nosrat, Robbie Jean . . . I love you all.

My mother-in-law, the historian Myra Rich, provided brilliant and sensitive reads and conversations. I feel so lucky to have her and Robert in my life. Marissa, mi hermana, you are the smartest, toughest person. You should run everything. My parents, Elizabeth and Salvador, are the best. Mom, you are the one person who I *know* would always read a draft as quickly as I wanted it read. And Dad, thank you for being the model for me that you are, and for your engagement with me and the work. I love you both.

Flora, you are only four months older than this project, and I am finally, finally finished. I am so glad you are still just beginning, and I look forward to watching you dance through this world. Orion, my courageous, independent kid—I wish I had your inner compass and quiet strength. Also, I would like to be able to draw as well as you did when you were three.

And Sarah, my love, my planet. Look at what we've built together. I love you.

Index

Page numbers in *italics* refer to photographs.

A NOTE ABOUT THE AUTHOR

Alexis Madrigal is a journalist who lives in Oakland, California. He is the cohost of KQED's current affairs show, *Forum*, and a contributing writer at *The Atlantic*, where he cofounded the COVID Tracking Project. Previously, he was the editor in chief of Fusion and a staff writer at *Wired*. Madrigal is the author of *Powering the Dream: The History and Promise of Green Technology*. He has been a visiting scholar at both UC Berkeley's School of Information and its Center for Science, Technology, Medicine and Society as well as an affiliate with the Berkman Klein Center for Internet and Society at Harvard University. He is the proprietor of *Oakland Garden Club*, a newsletter for people who like to think about plants. He was born in Mexico City, grew up in rural Washington State, and went to Harvard University.